Lecture Notes in Mathematics 1876

Horst Herrlich

Axiom of Choice

Author

Horst Herrlich

Department of Mathematics
University of Bremen
P.O. Box 33 04 40
28334 Bremen
Germany
e-mail: horst.herrlich@t-online.de

Library of Congress Control Number: 2006921740

Mathematics Subject Classification (2000): 03E25, 03E60, 03E65, 05C15, 06B10, 08B30, 18A40, 26A03, 28A20, 46A22, 54B10, 54B30, 54C35, 54D20, 54D30, 91A35

ISSN print edition: 0075-8434
ISSN electronic edition: 1617-9692
ISBN-10 3-540-30989-6 Springer Berlin Heidelberg New York
ISBN-13 978-3-540-30989-5 Springer Berlin Heidelberg New York

DOI 10.1007/11601562

Springer is a part of Springer Science+Business Media
springer.com

Printed in The Netherlands

Typesetting: by the author and TechBooks using a Springer LaTeX package
Cover design: *design & production* GmbH, Heidelberg

Printed on acid-free paper SPIN: 11601562 41/TechBooks 5 4 3 2 1 0

Dedicated in friendship
to George, Gerhard, and Lamar

It is a peculiar fact that all the transfinite axioms are deducible from a single one, the axiom of choice, — the most challenged axiom in the mathematical literature.

D. Hilbert (1926)

It is the great and ancient ***problem of existence*** *that underlies the whole controversy about the axiom of choice.*

W. Sierpiński (1958)

Wie die mathematische Analysis gewissermaßen eine einzige Symphonie des Unendlichen ist.

D. Hilbert (1926)

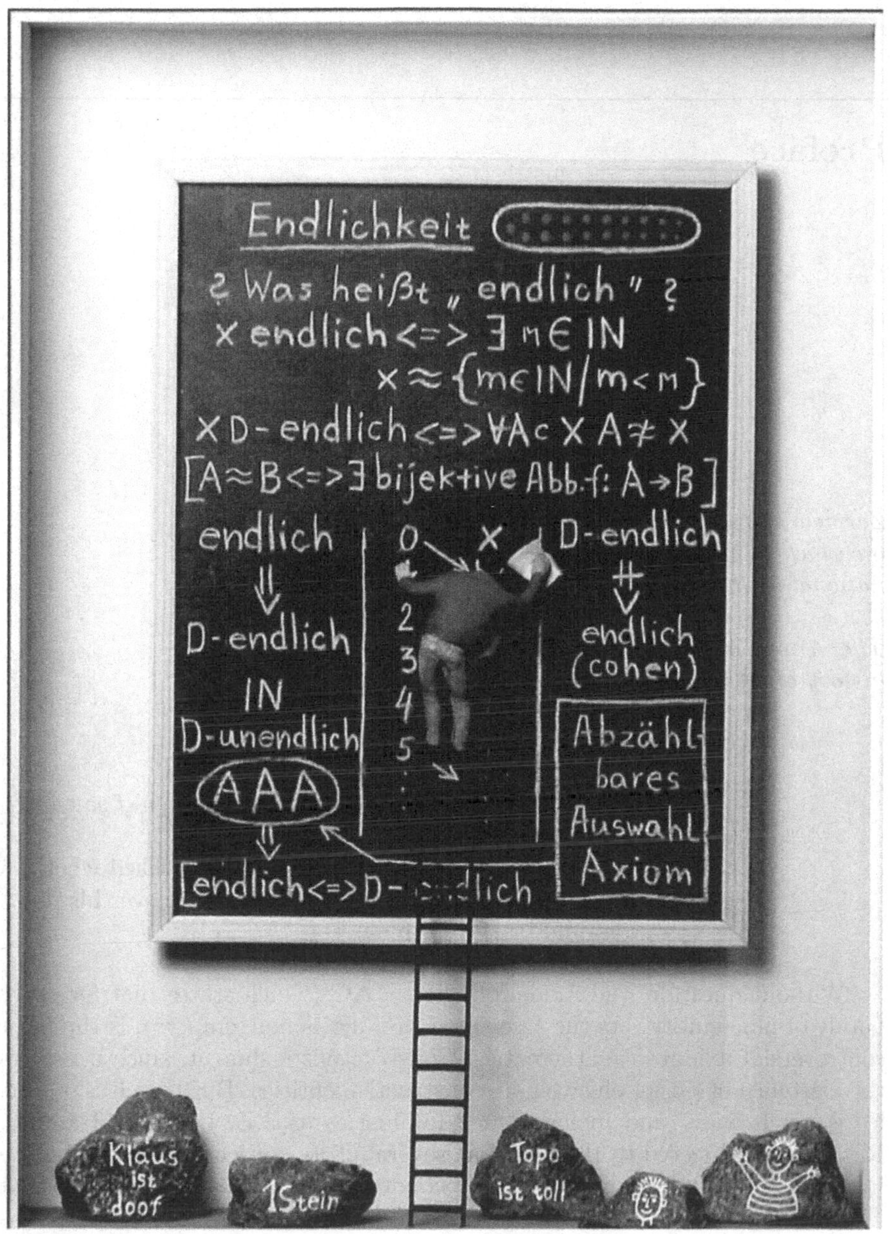

Der Mathematiker. Courtesy of the painter Volker Kühn (www.artinboxes.com).

Preface

Zermelo's proof, and especially the Axiom of Choice on which it was based, created a furor in the international mathematical community.

...

The Axiom of Choice has easily the most tortured history of all the set–theoretic axioms.

Penelope Maddy (Believing the axioms I)[1]

Of course not, but I am told it works even if you don't believe in it.

Niels Bohr (when asked whether he really believed a horseshoe hanging over his door would bring him luck).[2]

Without question, the Axiom of Choice, **AC** (which states that for every family of non–empty sets the associated product is non–empty[3]), is the most controversial axiom in mathematics. Constructivists shun it, since it asserts the existence of rather elusive non–constructive entities. But the class of critics is much wider and includes such luminaries as J.E. Littlewood and B. Russell who objected to the fact that several of its consequences such as the Banach–Tarski Paradox are extremely counterintuitive, and who claimed that *"reflection makes the intuition of its truth doubtful, analysing it into prejudices derived from the finite case"*[4], resp. that *"the apparent evidence of the*

1 [Mad88]

2 c. 1930. Cited from: The Oxford Dictionary of Modern Quotations. Second Edition with updated supplement. 2004.

3 cf. Definition 1.1.

4 [Lit26]

axiom tends to dissipate upon the influence of reflection"[5]. (See also the comments after Theorem 1.4.) Nevertheless, over the years the proponents of **AC** seemed to have won the debate, first of all due to the fact that disasters happen without **AC**: many beautiful theorems are no longer provable, and secondly, Gödel showed that **AC** is relatively consistent[6]. So **AC** could not be responsible for any antinomies which might emerge. This somewhat opportunistic attitude, sometimes supported by such arguments as *"Even if we knew that it was impossible ever to define a single member of a class, it would not of course follow that members of the class did not exist."*[7], led to the situation that in most modern textbooks **AC** is assumed to be valid indiscriminately. Still, these facts only show the usefulness of **AC** not its validity, and Lusin's verdict[8] *"For me the proof of a theorem by means of Zermelo's axiom is valuable only as an indication that it is useless to waste time on an exact proof of the falsity of the theorem in question"* is still shared at least by the constructivists. Unfortunately, our intuition is too hazy for considering **AC** to be *evidently* true or *evidently* false, as expressed whimsically by J.L. Bona: "*The Axiom of Choice is obviously true, the Well–Ordering Principle is obviously false; and who can tell about Zorn's Lemma*".[9]

Observe however that the distinction between the Axiom of Choice and the Well–Ordering Theorem is regarded by some, e.g. by H. Poincaré, as a serious one:

> *"The negative attitude of most intuitionists, because of the existential character of the axiom* [of choice], *will be stressed in Chapter IV. To be sure, there are a few exceptions, for the equivalence of the axiom to the well-ordering theorem (which is rejected by all intuitionists) depends, inter alia, on procedures of a supposedly impredicative character; hence the possibility exists of accepting the axiom but rejecting well-ordering as it involves impredicative procedures. This was the attitude of Poincaré."*[10]

When Paul Cohen demonstrated that the negation of **AC** is relatively consistent too[11], and when he created a method for constructing models of **ZF** (i.e., Zermelo–Fraenkel set theory without the Axiom of Choice) in which not only **AC** fails, but in which certain given substitutes of **AC** — either weakening **AC** or even contradicting **AC** — hold, he triggered *"the post Paul*

[5] [Rus11]
[6] [Goed39]
[7] [Hard06]
[8] Lusin 1926, cited after [Sie58, p. 95].
[9] [Sch97, p. 145]
[10] [FrBaLe73, p. 81]
[11] [Coh63/64]

Cohen set–theoretic renaissance"[12], and a vast literature emerged in which **AC** is not assumed; thus giving life to Sierpiński's program[13]:

> *"Still, apart from our being personally inclined to accept the axiom of choice, we must take into consideration, in any case, its role in the Set Theory and in the Calculus. On the one hand, since the axiom of choice has been questioned by some mathematicians, it is important to know which theorems are proved with its aid and to realize the exact point at which the proof has been based on the axiom of choice; for it has frequently happened that various authors had made use of the axiom of choice in their proofs without being aware of it. And after all, even if no one questioned the axiom of choice, it would not be without interest to investigate which proofs are based on it and which theorems can be proved without its aid.*
>
> ...
>
> *It is most desirable to distinguish between theorems which can be proved without the aid of the axiom of choice and those which we are not able to prove without the aid of this axiom.*
> *Analysing proofs based on the axiom of choice we can*
> 1. *ascertain that the proof in question makes use of a certain particular case of the axiom of choice,*
> 2. *determine the particular case of the axiom of choice which is sufficient for the proof of the theorem in question, and the case which is necessary for the proof ...*
> 3. *determine that particular case of the axiom of choice which is both necessary and sufficient for the proof of the theorem in question."*

This book is written in Sierpiński's spirit, but one more step will be added which occurred neither to Sierpiński nor to Lusin, but was made possible by Cohen's work that opened new doors for set theorists: *"Set theory entered its modern era in the early 1960's on the heels of Cohen's discovery of the method of forcing and Scott's discovery of the relationship between large cardinal axioms and constructible sets."*[14] Some striking theorems will be presented, that can be proved to be false in **ZFC** (i.e., Zermelo–Fraenkel set theory with the Axiom of Choice), but which hold in **ZF** provided **AC** is replaced by some (relatively consistent) alternative axiom.

This book is not written as a compendium, or a textbook, or a history of the subject — far more comprehensive treatments of specific aspects can be found in the list of Selected Books and Longer Articles. I hope, however, that this monograph might find its way into seminars. Its purpose is to whet the

---

[12] J.M. Plotkin in the Zentralblatt review Zbl. 0582.03033 of [RuRu85].
[13] [Sie58, p. 90 and 96] Cf. also [Sie18]
[14] [Kle77]

reader's appetite for studying the **ZF**–universe in its fullness, and not just its highly interesting but rather small **ZFC**–part. Mathematics is sometimes compared with a cathedral, the mathematicians being simultaneously its architects and its admirers. Why visit only one of it wings — the one built with the help of **AC**? Beauty and excitement can be found in other parts as well — and there is no law that prevents those who visit one of its parts from visiting other parts, too.

An attempt has been made to keep the material treated as simple and elementary as possible. In particular no special knowledge of axiomatic set theory is required. However, a certain mathematical maturity and a basic acquaintance with general topology will turn out to be helpful.

The sections can be studied more or less independently of each other. However, it is recommended not to skip any of the sections 2.1, 2.2, or 3.3 since they contain several basic definitions.

A treatise like this one does not come out of the blue. It rests on the work of many people. Acknowledgments are due and happily given:

- to all those mathematicians — living or dead — whose work I have cannibalized freely, most of all to Paul Howard and Jean Rubin for their wonderful book, Consequences of the Axiom of Choice,
- to those colleagues and friends whose curiosity, knowledge, and creativity provided ample inspiration, often leading to joint publications: Lamar Bentley, Norbert Brunner, Marcel Erné, Eraldo Giuli, Gonçalo Gutierres, Y.T. Rhineghost, George Strecker, Juris Steprāns, Eleftherios Tachtsis, and particularly Kyriakos Keremedis,
- to those who helped to unearth reprints: Lamar Bentley, Gerhard Preuss, and George Strecker,
- to those who read the text carefully to reduce the number of mistakes and to smoothen my imperfect English go very special thanks: Lamar Bentley, Kyriakos Keremedis, Eleftherios Tachtsis, Christoph Schubert, and particularly George Strecker,
- to Birgit Feddersen, my perfect secretary, who transformed my various crude versions of a manuscript miraculously into the present delightful shape,
- to Christoph Schubert for putting the final touches to the manuscript.

Let us end the preface with the following three quotes:

"Pudding and pie,"
Said Jane, "O, my!"
"Which would you rather?"
Said her father.
"Both," cried Jane,
Quite bold and plain.
Anonymous (ca. 1907)

The Axiom of Choice and its negation cannot coexist in one proof, but they can certainly coexist in one mind. It may be convenient to accept **AC** *on some days — e.g., for compactness arguments — and to accept some alternative reality, such as* **ZF** + **DC** + **BP**[15] *on other days — e.g., for thinking about complete metric spaces.*

E. Schechter (1997)[16]

So you see!
There's no end
To the things you might know,
Depending how far beyond Zebra you go!
Dr. Seuss (1955)[17]

15 **DC** is the *Principle of Dependent Choices*; see Definition 2.11.
BP stipulates that every subset of $\mathbb{R}$ has the *Baire property*, i.e., can be expressed as a symmetric difference of an open set and a meager set; see [Sch97].

16 [Sch97]

17 From *On Beyond Zebra.*

Contents

1

Origins

In 1904 the powder keg had been exploded through the match lighted by Zermelo.

A.A. Fraenkel, Y. Bar–Hillel, and A. Levy[1]

The Axiom of Choice (together with the Continuum Hypothesis) is probably the most interesting and most discussed axiom in mathematics after Euclid's Axiom of Parallels.

P. Bernays and A.A. Fraenkel[2]

In particular, since the continuum is virtually the set of all subsets of a denumerable set, solving the continuum problem possibly requires a more far–reaching characterization of the concept of subset than obtained above by the Axiom of Subsets and of Choice. What these two axioms furnish, may be separated by a deep abyss from what Cantor had in mind when speaking of s, viz. arbitrary multitudes of elements of s. One cannot expect to determine the number of the subsets of s before it is unambiguously settled what they are.

P. Bernays and A.A. Fraenkel[3]

1 [FrBaLe73, p. 84]
2 [BeFr58, p. 16]
3 [BeFr58, p. 26]

1.1 Hilbert's First Problem

Diese Disziplin ist die Mengenlehre, deren Schöpfer Georg Cantor war, ..., diese erscheint mir als die bewundernswerteste Blüte mathematischen Geistes und überhaupt eine der höchsten Leistungen rein verstandesmäßiger menschlicher Tätigkeit

...

Aus dem Paradies, das Cantor uns geschaffen, soll uns niemand vertreiben können.[4]

D. Hilbert

Everyone agrees that, whether or not one believes that set theory refers to an existing reality, there is a beauty in its simplicity and in its scope.

P. Cohen[5]

Cantor, after creating set theory, left two famous conjectures:

- the **Continuum Hypothesis** (**CH**), stating that every infinite subset of the reals is either countable or has the same cardinality as $\mathbb{R}$ itself, and
- the **Well–Order Theorem** (**WOT**), stating that every set can be well–ordered,

the latter of these he originally regarded as a self–evident law of thought[6], but later convinced himself that it required a proof.

Hilbert considered these problems as so fundamental, in particular for our understanding of the concept of the *continuum*, that in his 1900 Paris lecture[7] — without question the most influential mathematical lecture ever given — he formulated the conjunction of Cantor's two conjectures as the first of his famous problems:

---

[4] This field is the theory of sets, whose creator was Georg Cantor, ..., this appears to me as the most marvelous fruit of the mathematical mind, indeed as one of the highest achievements of purely rational human activities. ... Nobody is to banish us from the paradise created by Cantor. [Hil26].

[5] [Coh2002]

[6] "Daß es immer möglich ist, jede *wohldefinierte* Menge in die *Form* einer *wohlgeordneten* Menge zu bringen, auf dieses, wie mir scheint, grundlegende und folgenreiche, durch seine Allgemeingültigkeit besonders merkwürdige Denkgesetz werde ich in einer späteren Abhandlung zurückkommen." Cantor (1883)

[7] [Hil1900]

Hilbert's First Problem

1. CANTOR's Problem of the Cardinal Number of the Continuum

Two systems, i.e., two assemblages of ordinary real numbers or points, are said to be (according to Cantor) equivalent or of equal *cardinal number*, if they can be brought into a relation to one another such that to every number of the one assemblage corresponds one and only one definite number of the other. The investigations of Cantor on such assemblages of points suggest a very plausible theorem, which nevertheless, in spite of the most strenuous efforts, no one has succeeded in proving. This is the theorem:
Every system of infinitely many real numbers, i.e., every assemblage of numbers (or points), is either equivalent to the assemblage of natural integers, 1, 2, 3, ... or to the assemblage of all real numbers and therefore to the continuum, that is, to the points of a line; *as regards equivalence there are, therefore, only two assemblages of numbers, the countable assemblage and the continuum.*
From this theorem it would follow at once that the continuum has the next cardinal number beyond that of the countable assemblage; the proof of this theorem would, therefore, form a new bridge between the countable assemblage and the continuum.

Let me mention another very remarkable statement of Cantor's which stands in the closest connection with the theorem mentioned and which, perhaps, offers the key to its proof. Any system of real numbers is said to be ordered, if for every two numbers of the system it is determined which one is the earlier and which the later, and if at the same time this determination is of such a kind that, if a is before b and b is before c, then a always comes before c. The natural arrangement of numbers of a system is defined to be that in which the smaller precedes the larger. But there are, as is easily seen infinitely many other ways in which the numbers of a system may be arranged.
If we think of a definite arrangement of numbers and select from them a particular system of these numbers, a so–called partial system or assemblage, this partial system will also prove to be ordered. Now Cantor considers a particular kind of ordered assemblage which he designates as a well ordered assemblage and which is characterized in this way, that not only in the assemblage itself but also in every partial assemblage there exists a first number. The system of integers 1, 2, 3, ... in their natural order is evidently a well ordered assemblage. On the other hand the system of all real numbers, i.e., the continuum in its natural order, is evidently not well ordered. For, if we think of points of a segment of a straight line, with its initial point excluded, as our partial assemblage, it will have no first element.
The question now arises whether the totality of all numbers may not be arranged in another manner so that every partial assemblage may have a first element, i.e., whether the continuum cannot be considered as a well ordered assemblage — a question which Cantor thinks must be answered in the affirmative. It appears to me most desirable to obtain a direct proof of this remarkable statement of Cantor's, perhaps by actually giving an arrangement of numbers such that in every partial system a first number can be pointed out.

Let us pause for a moment and try to reconstruct the ideas that led to Cantor's second conjecture, restricted to the set of real numbers, and to analyze the relations between the two conjectures. We may try to well–order $\mathbb{R}$ as follows:

Pick a first element x_1, then a second x_2, then a third x_3, and so on, picking x_n for each natural number n different from the ones picked previously. Since $\mathbb{R}$ is uncountable, the x_n's do not exhaust $\mathbb{R}$. So, after infinitely many steps, we have to continue by picking new elements $x_\omega, x_{\omega+1}, \ldots, x_\alpha, \ldots$ where α runs through the collection of all countable ordinals. Here a serious problem arises: Not only does the ordered set of all countable ordinals have a very complicated structure, if considered in **ZFC**, but — even worse — its structure is not determined by **ZF**. There are several models for it with quite different properties[8]. Our intuition fails to help us decide on the "right" one — if there is such a thing at all. Our procedure starts to get somewhat nebulous, even obscure. But suppose that we can do it anyway. What then? There may still be real numbers left unpicked — and, if the first conjecture is false, there will be. So at least in the latter case, but in general also if **CH** holds, we will have to continue with our picking process. How far do we have to continue? This will depend on the size of $\mathbb{R}$, so here Cantor's first conjecture plays a vital part. Modern set theorists consider the size of $\mathbb{R}$ to be bigger than $\aleph_1$.

Gödel (1947)[9] conjectured it to be $\aleph_2$ and wrote:

> *"Therefore one may on good reason suspect that the role of the continuum problem in set theory will be this; that it will finally lead to the discovery of new axioms which will make it possible to disprove Cantor's conjecture."*

This is indeed what Woodin and other leading set theorists are trying to do now. Cohen (1966)[10] however went even further:

> *"A point of view which the author feels may eventually come to be accepted is that* ***CH*** *is* ***obviously*** *false.*
> ...
> *This point of view regards* [the continuum] *as an incredibly rich set given to us by one bold new axiom* [the Power Set Axiom; recall that $|\mathbb{R}| = |\mathcal{P}(\mathbb{N})|$], *which can never be approached by any piecemeal process of construction."*

---

8 *"If there be two of these postulates neither of which leads to contradiction* [which is indeed the case, provided **ZF** is consistent], *then there are corresponding to them two distinct self–consistent second ordinal classes, just as Euclidean geometry and Lobachevskian geometry are distinct self–consistent geometries, with, however, the difference, that the two second ordinal classes are incapable of existing together in the same universe of discourse."* [Chu27, p. 187–188]

9 [Goed47]

10 [Coh66]

In fact the independence results of Cohen and his successors show that any $\aleph$ which is not a countable sum of smaller $\aleph$'s can be made to be the cardinality of the continuum $\mathbb{R}$.

So, can we well–order the reals by means of the above procedure? Unlikely!

Let us return to Hilbert's First Problem. In 1904 Zermelo provided a positive solution to its second part by showing in a 3–page paper by means of a new axiom, the Axiom of Choice, that every set can be well–ordered.[11] This result brought him instant fame, a professorship in Göttingen one year later, and — created a big controversy about the validity of the Axiom of Choice.

Definition 1.1. **AC**, *the* Axiom of Choice, *states that for each family* $(X_i)_{i\in I}$ *of non–empty sets* X_i, *the product set* $\prod\limits_{i\in I} X_i$ *is non–empty.*[12]

Fact 1.2. For every set X there exists an ordinal α with $|\alpha| \not\leq |X|$.

This fact follows from the observation that there is only a set of possible well–orderings of X. We omit the technical details of a proof in **ZF**. But we mention, for later use, that the above fact guarantees the existence of a smallest such α. This must automatically be a cardinal, thus in the infinite case an $\aleph$. It has a name:

Definition 1.3. *For any infinite set* X, *the smallest* $\aleph$ *with* $\aleph \not\leq |X|$ *is called the* Hartogs–number[13] *of* X.

Theorem 1.4. *Equivalent are:*

1. **AC**.
2. **WOT**.

Proof. **(1) ⇒ (2)** Let X be an infinite set (the result is obvious for finite sets). By (1) the family $\mathcal{P}_0X$ of all non–empty subsets of X, indexed by itself, has a non–empty product, i.e., there exists a map $f\colon \mathcal{P}_0X \to X$ with $f(A) \in A$ for each $A \in \mathcal{P}_0X$. Let $\aleph = \{\alpha \in \mathrm{Ord} \mid \alpha < \aleph\}$ be the Hartogs–number of X and define, via transfinite recursion, a function $g\colon \aleph \to X \cup \{\infty\}$ by

$$g(\alpha) = \begin{cases} f(X \setminus \{g(\beta) \mid \beta < \alpha\}), & \text{if } X \neq \{g(\beta) \mid \beta < \alpha\} \\ \infty, & \text{otherwise} \end{cases}.$$

Since $\aleph \not\leq |X|$ there exists some α with $g(\alpha) = \infty$. If $\gamma = \min\{\alpha < \aleph \mid g(\alpha) = \infty\}$, then the restriction of g to a map from $\gamma = \{\alpha \in \mathrm{Ord} \mid \alpha < \gamma\}$ to X is a bijection. Thus with γ also X is well–orderable.

[11] [Zer04]. See also [Zer08] and [Zer08a].

[12] Note that the elements of the product set $\prod\limits_{i\in I} X_i$ are *choice functions* $(x_i)_{i\in I}$, i.e., functions $x\colon I \to \bigcup\limits_{i\in I} X_i$, satisfying $x(i) = x_i \in X_i$ for each $i \in I$.

[13] $\aleph = \sup\{\alpha \in \mathrm{Ord} \mid |\alpha| \leq |X|\}$. Cf.[Har15].

(2) ⇒ (1) Let $(X_i)_{i \in I}$ be a family of non–empty sets. Well–order the union of all the X_i's, and choose in X_i its smallest member x_i. Then $(x_i) \in \prod_{i \in I} X_i$.

Observe that the above proof of **AC** ⇒ **WOT** is nothing but a formalization of the naive procedure sketched earlier, where the "picking" of elements in X is done once and for all by our marvelous and mysterious apparatus, the Axiom of Choice.

Why the objections?

First of all constructivists, particularly intuitionists, deny the existence of all things that cannot be "constructed". E.g.:

"A formal system in which $\exists x G(x)$ is provable, but which provides no method for finding the x in question, is one in which the existential quantifier fails to fulfill its intended function."

R.L. Goodstein (1968)[14]

However skepticism towards the validity of **AC** is not restricted to constructivists, whose idea of "existence" is decidedly narrower than that of the majority of mathematicians, leading them to such extremes as the abolishment of the *law of excluded middle*[15], the assertion that all real functions are continuous, and the claim that there exist real numbers a and b, different from 0, such that the function $f\colon \mathbb{R} \to \mathbb{R}$, defined by $f(x) = a \cdot x + b$, has no root[16].

Consider, for example, the following statements:

*"It [**AC**] may be true but it lacks obviousness, and the conclusions drawn from it are astonishing. In these circumstances I think it would be well to abstain from using it.*

...

The apparent evidence of the axiom tends to dissipate upon the influence of reflection.

...

In the end one ceases to understand what it means.

...

In my opinion there is no reason whatsoever to believe the truth of the axiom."

B. Russell (1911)[17]

---

14 [Goo68]

15 W.V.O. Quine comments on this: "The doctrine ... has led its devotees to such quixotic extremes as that of challenging the method of proof by *reductio ad absurdum* — a challenge in which I sense a *reductio ad absurdum* of the doctrine itself." Note that **AC** implies the law of excluded middle (see [GoMy78]).

16 [Hey56]

17 [Rus11]

> *"The axiom of choice has many elegant consequences, but that is an argument for its mathematical interest, not for its truth.*
>
> *...*
>
> *The formal consistency of making this assumption* [validity of the Axiom of Strong Choice] *can hardly be doubted, but it ascribes to us abilities which I for one am not aware of possessing.*
>
> *...*
>
> *It is dangerous to claim the existence of an object one cannot describe: "Whereof one cannot speak thereof one must be silent", to hijack a slogan (Wittgenstein 1922)."*
>
> M.D. Potter (1990)[18]

Other mathematicians had difficulties deciding for or against the axiom, e.g., van der Waerden:

> *"In 1930, van der Waerden published his* **Modern Algebra**, *detailing the exciting new applications of the axiom. The book was very influential, providing Zorn and Teichmüller with a proving ground for their versions of choice, but van der Waerden's Dutch colleagues persuaded him to abandon the axiom in the second edition of 1937. He did so, but the resulting limited version of abstract algebra brought such a strong protest from his fellow algebraists that he was moved to reinstate the axiom and all its consequences in the third edition of 1950."*
>
> P. Maddy (1988)[19]

After Gödel (1938) proved the relative consistency of the Axiom of Choice by constructing within a given model of **ZF** a model of **ZFC**, the proponents of **AC** gained ground. Most modern textbooks take **AC** for granted and the vast majority of mathematicians use **AC** freely. However, after Cohen (1963) proved the relative consistency of the negation of **AC** and, moreover, provided a method, called *forcing*, for producing a plethora of models of **ZF** that have or fail to have a wide range of specified properties, a growing number of mathematicians started to investigate the **ZF** world by substituting **AC** by a variety of possible alternatives, sometimes just by weakening **AC** and sometimes by replacing **AC** by axioms that contradict it.

All this work demonstrates how useful or convenient such axioms as **AC** and its possible alternatives are. But the question of the truth of **AC** is not touched, and Hilbert's First Problem remains unanswered. It is conceivable, even likely, that it will never be solved[20], despite Hilbert's optimistic slogan expressed in his Paris lecture: "in mathematics there is no *ignorabimus*."

---

18 [Pot90]

19 [Mad88]

20 "And whether, in particular, Zermelo's axiom is true or false is a question, which, while more fundamental matters are in doubt, is very likely to remain unanswered." Russell (1907)

Our inability to decide on such basic questions of set theory, as those formulated in Hilbert's First Problem, leads inevitably to the conclusion formulated by Mostowski[21] that

> *"Probably we shall have in the future essentially different intuitive notions of sets just as we have different notions of space."*

In some of these **ZF**–worlds Hilbert's First Problem will have a positive solution, in others a negative one.

Exercises to Section 1.1:

E 1. Show that, for a set X, the following conditions are equivalent:
(1) X is well–orderable.
(2) There exists a function $f\colon \mathcal{P}_0 X \to X$ with $f(A) \in A$ for each $A \in \mathcal{P}_0 X$.
(3) $\mathbf{AC}(X)$, i.e., $\prod_{i\in I} X_i \neq \emptyset$ for each family $(X_i)_{i\in I}$ of non–empty subsets X_i of X.
[Hint: Analyze the proof of Theorem 1.4.]

E 2. Let $(X_i)_{i\in I}$ be a family of non–empty sets. Show that in each of the following situations, $\prod_{i\in I} X_i \neq \emptyset$:
(1) I is finite.
(2) The X_i's are well–ordered sets.
(3) The X_i's are finite linearly ordered sets.
(4) The X_i's are finite subsets of $\mathbb{R}$.
(5) The X_i's are closed subsets of $\mathbb{R}$.
(6) The X_i's are open subsets of $\mathbb{R}$.
(7) $\bigcup_{i\in I} X_i$ is well–orderable.
(8) $\bigcup_{i\in I} X_i$ is linearly orderable and each X_i is finite.

E 3. [22] Construct a set X with the following properties:
(1) X is well–orderable.
(2) $|X| \not\leq |\mathbb{R}|$.
(3) $|\mathbb{R}| \not\leq |X|$.
(4) $|X| = |\mathbb{R}|$ iff $\mathbb{R}$ is well–orderable.
[Hint: Use Fact 1.2.]

21 [Mos67]
22 [Sie21]

2

Choice Principles

2.1 Some Equivalents to the Axiom of Choice

We believe that the ZF–axioms describe in a correct way our intuitive contemplations concerning the notion of set. The axiom of choice (AC) is intuitively not so clear as the other ZF–axioms are, but we have learned to use it because it seems to be indispensable in proving mathematical theorems. On the other hand the (AC) has "strange" consequences, such as "every set can be well–ordered" and we are unable to "imagine" a well–ordering of the set of real numbers.

U. Felgner[1]

Once this method for unveiling the truth had been discovered by him, he found it indispensable.

Ivan Olbracht[2]

Let us start with some familiar observations:

Proposition 2.1. *Equivalent are:*

1. **AC**.
2. *For every family $(X_i)_{i\in I}$ of non–empty pairwise disjoint sets there exists a set Y with $|Y \cap X_i| = 1$ for each $i \in I$.*

Proof. Exercise.

Observe that the above condition (2) has been introduced under the name *multiplicative axiom* by Russell, since it allowed the definition of arbitrary products $\prod_{i\in I} a_i$ of cardinal numbers a_i.

---

1 [Fel71]

2 From: *The Good Judge* (Dobrý Soudce).

Though nowadays the proof of the above proposition is a trivial exercise, this was not the case when Russell submitted his paper in 1905 as the following quote[3] reveals[4].

> *"This axiom is more special than Zermelo's axiom. It can be deduced from Zermelo's axiom; but the converse deduction, though it may turn out to be possible, has not yet, so far as I know, been effected. I shall call this the* multiplicative *axiom."*

Theorem 2.2. *Equivalent are:*

1. **AC**.
2. **Hausdorff's Maximal Chain Condition:** *Each partially ordered set contains a maximal chain*[5].
3. **Zorn's Lemma:** *If in a partially ordered set X each chain has an upper bound, then X has a maximal element.*
4. **Teichmüller–Tukey Lemma:** *If a non-empty subcollection $\mathfrak{U}$ of $\mathcal{P}X$ is of finite character*[6]*, then $\mathfrak{U}$ contains a maximal element w.r.t. the inclusion-order.*
5. *Each preordered set contains a maximal antichain*[7].

Proof. **(1) ⇒ (2)** Let X be a partially ordered set without a maximal chain. Then for each chain K in X the set

$$C(K) = \{x \in (X \setminus K) \mid K \cup \{x\} \text{ is a chain}\}$$

is non-empty. By (1), there exists a map $f\colon \mathcal{P}_0X \to X$ with $f(A) \in A$ for each $A \in \mathcal{P}_0X$. Let $\aleph$ be the Hartogs-number of X. Define, via transfinite recursion, a map $g\colon \aleph \to X$ by

$$g(\alpha) = f(C\{g(\beta) \mid \beta < \alpha\}).$$

Then g is injective, a contradiction.

(2) ⇒ (3) Let X be a partially ordered set, satisfying the premise of Zorn's Lemma. By (2), X has a maximal chain K. Let x be an upper bound of K. Then x is a maximal element of X.

(3) ⇒ (4) Let $\mathfrak{U}$ be a non-empty subset of $\mathcal{P}X$ that is of finite character. Then every chain in $\mathfrak{U}$, ordered by inclusion, has an upper bound, its union. Thus, by (3), $\mathfrak{U}$ has a maximal element.

---

[3] [Rus07]

[4] G.H. Hardy made a similar statement: "*... but whether or not the latter imply the former has not yet been decided.*" [Hard06]

[5] X is called a *chain* iff $x \leq y$ or $y \leq x$ for any x and y in X.

[6] $\mathfrak{U}$ is called *of finite character* provided that $A \in \mathfrak{U}$ iff every finite subset of A belongs to $\mathfrak{U}$.

[7] X is called an *antichain* iff $x \leq y$ implies $x = y$ for any x and y in X.

(4) ⇒ (5) Let X be a preordered set. The collection $\mathfrak{U}$ of all antichains in X is of finite character and non–empty (since $\emptyset \in \mathfrak{U}$). Thus, by (4), there exists a maximal element of $\mathfrak{U}$.

(5) ⇒ (1) Let $(X_i)_{i\in I}$ be a family of non–empty sets. Preorder the set $X = \{(x,i) \mid i \in I \text{ and } x \in X_i\}$ by

$$(x,i) \leq (y,j) \Leftrightarrow i = j.$$

Then a maximal antichain K of X has the form $K = \{(x_i, i) \mid i \in I\}$ where, for each $i \in I$, x_i is a distinguished element of X_i. Thus $\prod_{i\in I} X_i \neq \emptyset$.

Condition (5) of the above theorem can be weakened by replacing preorders by partial orders. That the resulting condition is in fact equivalent to (5) and thus to **AC** is not obvious. To show this and some further equivalences to **AC** we need the following lemma that we present without proof, since the latter would require greater familiarity with the axioms of **ZF** than we presuppose for our text; in particular it rests crucially on the axiom of foundation:

Lemma 2.3. [8] *If for every well–orderable set X the power set $\mathcal{P}X$ is also well–orderable, then every set is well–orderable.*

Theorem 2.4. [9] *Equivalent are:*

1. **AC**.
2. **The Axiom of Multiple Choice (AMC):** *For every family $(X_i)_{i\in I}$ of non–empty sets there exists a family $(F_i)_{i\in I}$ of non–empty, finite sets F_i with $F_i \subseteq X_i$ for each $i \in I$.*
3. **Kurepa's Maximal Antichain Condition:** *Each partially ordered set has a maximal antichain.*
4. *Every chain can be well–ordered.*
5. *The power set $\mathcal{P}X$ of each well–orderable set X is well–orderable.*

Proof. **(1) ⇒ (2)** Trivial.

(2) ⇒ (3) Let X be a partially ordered set. By (2), there exists a map $f\colon \mathcal{P}_0X \to \mathcal{P}_{\text{fin}}X$ from $\mathcal{P}_0X$ into the set $\mathcal{P}_{\text{fin}}X$ of all non–empty, finite subsets of X with $f(A) \subseteq A$ for each $A \in \mathcal{P}_0X$. Let $g(A)$ be the set of all minimal elements of $f(A)$. Then each $g(A)$ is a non–empty antichain with $g(A) \subseteq A$. Define further, for each antichain K of X,

$$A(K) = \{x \in (X \setminus K) \mid K \cup \{x\} \text{ is an antichain}\}.$$

If X would have no maximal antichain, then all the $A(K)$'s would be non–empty. Thus we could define an injective map $h\colon \aleph \to \mathcal{P}X$ from the Hartogs–number $\aleph$ of $\mathcal{P}X$ into $\mathcal{P}X$ via transfinite recursion by

---
[8] [Rub60], [FeJe73].
[9] [Rub60], [FeJe73].

$$h(\alpha) = \bigcup_{\beta<\alpha} h(\beta) \cup g\left(A\left(\bigcup_{\beta<\alpha} h(\beta)\right)\right).$$

Contradiction.

(3) ⇒ (4) Let X be a chain. Define a partially ordered set Y by $Y = \{(A, a) \mid A \subseteq X \text{ and } a \in A\}$ and $(A, a) \leq (B, b)$ iff ($A = B$ and $a \leq b$ in X). Then an antichain K of Y has the form

$$K = \{(A, a(A)) \mid A \in \mathcal{P}_0X\}$$

where each $a(A)$ is a distinguished element of A. Thus X is well–orderable by Exercises to Section 1.1, E 1.

(4) ⇒ (5) Let X be well–ordered. Then $\mathcal{P}X$ is linearly ordered by

$$A < B \Leftrightarrow \exists a \in (A \setminus B) \;\; \forall b \in (B \setminus A) \;\; a < b.$$

Thus, by (4), $\mathcal{P}X$ is well–orderable.

(5) ⇒ (1) Lemma 2.3 and Theorem 1.4.

Exercises to Section 2.1:

E 1. [10] Show (without using Theorem 2.4) that **AC** is equivalent to the conjunction of the following 2 conditions:

a) **OAC**: For every family $(X_i)_{i\in I}$ of non–empty sets there exists a family $(F_i)_{i\in I}$ of finite sets with $|F_i|$ odd and $F_i \subseteq X_i$.

b) **EAC**: For every family $(X_i)_{i\in I}$ of sets X_i with $|X_i| \geq 2$ there exists a family $(F_i)_{i\in I}$ of non–empty, finite sets F_i with $|F_i|$ even and $F_i \subseteq X_i$.

E 2. Show that **AC** is equivalent to the condition

(*) Each set (of sets) contains a subset that is maximal w.r.t. the property that its members are pairwise disjoint.

[Hint: For each family $(X_i)_{i\in I}$ of pairwise disjoint non–empty sets, consider the set

$$\{\{(0, x),\ (1, X_i)\} \mid i \in I \text{ and } x \in X_i\}.]$$

E 3. Show that **AC** is equivalent to the following version of Zorn's Lemma: If in an ordered set X a subset A has an upper bound in X whenever each pair of elements of A has an upper bound in A, then X has a maximal element.

[10] [Ker96]

E 4. Show the equivalence of:

a) **AC**.

b) Every surjection $f\colon X \to Y$ is a *retraction*, i.e., there exists a map $g\colon Y \to X$ with $f \circ g = \mathrm{id}_Y$ (= the identity map on Y).

c) Every set X is *projective*, i.e., for each map $f\colon X \to Y$ and each surjection $g\colon Z \to Y$ there exists a map $k\colon X \to Z$ with $f = g \circ k$.

d) Every set is a subset of some projective set (i.e., there exist arbitrary large projective sets with respect to the order $\leq$ of cardinals).

e) For every relation φ on a set X satisfying $\forall x \in X \ \exists y \in X \ x\varphi y$, there exists a function $f\colon X \to X$ with $x\varphi f(x)$ for each $x \in X$.

[Hint: For (e) $\Rightarrow$ (a) and a family $(X_i)_{i\in I}$ of non–empty sets, consider (X, φ) with $X = I \uplus \bigcup_{i\in I} X_i$ and $\varphi = \{(x,x) \mid x \in \bigcup_{i\in I} X_i\} \cup \{(i,x) \mid i \in I$ and $x \in X_i\}$.]

E 5. Show that **AC** holds iff for each family $(X_i)_{i\in I}$ there exists a family $(Y_i)_{i\in I}$ of pairwise disjoint subsets Y_i of X_i with $\bigcup_{i\in I} Y_i = \bigcup_{i\in I} X_i$.

2.2 Some Concepts Related to the Axiom of Choice

The last concept [**AC**] *seems to me to be entirely devoid of sense. As regards a denumerable infinity of choices, they cannot, of course, all be performed, but we can at least indicate such a procedure that, if we establish it beforehand, we may be sure that each choice will be made within a finite period of time; therefore, if two given systems of choice are different, we are sure to notice this after a finite number of operations. When an infinite number of choices is not denumerable, it is impossible to imagine a way of defining it, i.e., distinguishing it from an analogous infinite number of choices; thus it is impossible to regard it as a mathematical creation which can be introduced in arguments.*

E. Borel[11]

The notion that there is nothing special about countability, which is most pervasive in the works of Bourbaki, makes assuming only that every ***countable*** *set has a choice function seem a mite perverse.*

M.D. Potter[12]

Some mathematicians, who reject the Axiom of Choice, still accept some of its weaker forms, e.g., **CC**, the Axiom of Countable Choice, or **DC**, the Principle of Dependent Choices, while others advocate either all or nothing.

In this section we will formulate several axioms, related to **AC**, and discuss their mutual relations. Natural ways to weaken **AC** are obtained by the following procedures:

---

[11] [Bor14], quoted from [Sie58].

[12] [Pot90]

(A) Restrict the size of the indexing set I.
(B) Restrict the nature of the single X_i's.
(C) Replace the stipulation that one can exhibit simultaneously in each X_i a distinguished point x_i, equivalently: a one–element subset $\{x_i\}$, by some weaker requirement.

Procedure (A) gives:

Definition 2.5. **CC**, *the* Axiom of Countable Choice, *states that for each sequence* $(X_n)_{n\in\mathbb{N}}$ *of non–empty sets* X_n, *the product set* $\prod_{n\in\mathbb{N}} X_n$ *is non–empty.*

Procedure (B) gives:

Definition 2.6. *1.* **AC**(fin) *states that for each family* $(X_i)_{i\in I}$ *of non–empty finite sets* X_i, *the product set* $\prod_{i\in I} X_i$ *is non–empty.*

2. **AC**(n), *for* $n \in \mathbb{N}$, *states that for each family* $(X_i)_{i\in I}$ *of* n*–element sets, the product* $\prod_{i\in I} X_i$ *is non–empty.*

Procedure (C) gives:

Definition 2.7. **AMC**, *the* Axiom of Multiple Choice, *states that for each family* $(X_i)_{i\in I}$ *of non–empty sets* X_i, *there exists a family* $(F_i)_{i\in I}$ *of non–empty finite subsets* F_i *of* X_i.

Definition 2.8. **KW**, *the* Kinna–Wagner Selection Principle, *states that for each family* $(X_i)_{i\in I}$ *of at least 2–element sets* X_i, *there exists a family* $(Y_i)_{i\in I}$ *of non–empty proper subsets* Y_i *of* X_i.

Combining the procedures (A) and (B) one obtains:

Definition 2.9. *1.* **CC**$(\mathbb{R})$ *states that for each sequence* $(X_n)_{n\in\mathbb{N}}$ *of non–empty subsets* X_n *of* $\mathbb{R}$, *the product set* $\prod_{n\in\mathbb{N}} X_n$ *is non–empty.*

2. **CC**$(\mathbb{Z})$ *states that for each sequence* $((X_n, \leq_n))_{n\in\mathbb{N}}$, *each* $(X_n, \leq_n)$ *being order–isomorphic to the ordered set of integers, the product set* $\prod_{n\in\mathbb{N}} X_n$ *is non–empty.*

3. **CC**(fin) *states that for each sequence* $(X_n)_{n\in\mathbb{N}}$ *of non–empty, finite sets, the product set* $\prod_{n\in\mathbb{N}} X_n$ *is non–empty.*

4. **CC**(n), *for* $n \in \mathbb{N}^+$, *states that for each sequence* $(X_n)_{n\in\mathbb{N}}$ *of* n*–element sets, the product set* $\prod_{n\in\mathbb{N}} X_n$ *is non–empty.*

Combining the procedures (B) and (C) one obtains:

Definition 2.10. **CMC**, *the* Axiom of Countable Multiple Choice, *states that for each sequence* $(X_n)_{n\in\mathbb{N}}$ *of non–empty sets* X_n, *there exists a sequence* $(F_n)_{n\in\mathbb{N}}$ *of non–empty finite subsets* F_n *of* X_n.

Closely related to **CC** are the following two axioms:

Definition 2.11. *1.* **DC**, *the* Principle of Dependent Choices, *states that for every pair* (X, ϱ), *where* X *is a non-empty set and* ϱ *is a relation on* X *such that*

$$\text{for each} \quad x \in X \quad \text{there exists} \quad y \in X \quad \text{with} \quad x \varrho y,$$

there exists a sequence (x_n) *in* X *with* $x_n \varrho x_{n+1}$ *for each* $n \in \mathbb{N}$.

2. **PCC**, *the* Axiom of Partial Countable Choice, *states that for each sequence* $(X_n)_{n\in\mathbb{N}}$ *of non-empty sets* X_n, *there exists an infinite subset* M *of* $\mathbb{N}$ *with* $\prod_{m\in M} X_m \neq \emptyset$.

Theorem 2.12. *1.* **AC** $\Rightarrow$ **DC**.
2. **DC** $\Rightarrow$ **CC**.
3. **CC** $\Leftrightarrow$ **PCC**.

Proof. **(1)** Let (X, ϱ) be as specified in **DC**. Then, for each $x \in X$, the set $S_x = \{y \in X \mid x\varrho y\}$ is non-empty. Thus, by **AC**, there exists an element $(s_x)_{x\in X}$ in $\prod_{x\in X} S_x$. Choose an arbitrary $x_0 \in X$ and define, via recursion, a sequence (x_n) by: $x_{n+1} = s_{x_n}$.

Then (x_n) has the desired property.

(2) Let $(X_n)_{n\in\mathbb{N}^+}$ be a sequence of non-empty sets X_n. Define $Y_n = \prod_{m\leq n} X_m$ and
$Y = \bigcup_{n\in\mathbb{N}^+} Y_n$. Let ϱ be the relation defined on Y by:

$$(x_1 \ldots, x_n)\varrho(z_1, \ldots, z_m) \Leftrightarrow (m = n + 1 \text{ and } x_i = z_i \text{ for } i = 1, \ldots, n).$$

By **DC**, there exists a sequence (y_n) in Y with $y_n \varrho y_{n+1}$ for each $n \in \mathbb{N}^+$. Assume for simplicity that $y_1 = (x_1)$, (cf. Exercise E 3). Then each y_n has the form $(x_1^n, x_2^n, \ldots, x_n^n) \in \prod_{m\leq n} X_m$. Thus $(x_n^n)_{n\in\mathbb{N}^+}$ is an element of $\prod_{n\in\mathbb{N}^+} X_n$.

(3) Obviously **CC** implies **PCC**. For the converse consider a sequence $(X_n)_{n\in\mathbb{N}^+}$ of non-empty sets. Define $Y_n = \prod_{m\leq n} X_m$. Then $(Y_n)_{n\in\mathbb{N}^+}$ is a sequence of non-empty sets. By **PCC** there exists an infinite subset M of $\mathbb{N}$ and an element $(y_m)_{m\in M}$ of $\prod_{m\in M} Y_m$. Then $y_m = (x_1^m, \ldots, x_m^m) \in \prod_{k\leq m} X_k$. For each $n \in \mathbb{N}^+$ let $m(n) = \min\{m \in M \mid n \leq m\}$. Then $(x_n^{m(n)})_{n\in\mathbb{N}^+}$ is an element of $\prod_{n\in\mathbb{N}^+} X_n$.

There are many other types of weak choice principles[13]. Next, we present a few of these, that play a prominent part in subsequent sections of this treatise.

[13] In [HoRu98] there is a list of 383 such "forms".

Definition 2.13. *1.* **Fin** *states that every infinite set X is Dedekind–infinite*[14], *i.e., allows an injection* $\mathbb{N} \to X$.
2. **Fin**($\mathbb{R}$) *states that every infinite subset of* $\mathbb{R}$ *is D–infinite.*
3. **Fin**(lin) *states that every infinite, linearly ordered set is D–infinite.*

Theorem 2.14. *1.* **CC** $\Rightarrow$ **Fin**.
2. **Fin** $\Rightarrow$ **CC**(fin).

Proof. **(1)** See Proposition 4.13.
(2) Let $(X_n)_{n\in\mathbb{N}}$ be a sequence of non–empty, finite sets. Then $X = \bigcup_{n\in\mathbb{N}}(X_n \times \{n\})$ is an infinite set, thus, by **Fin**, D–infinite. Let $f\colon \mathbb{N} \to X$ be an injection. Since each X_n is finite, the set $M = \{n \in \mathbb{N} \mid f[\mathbb{N}] \cap (X_n \times \{n\}) \neq \emptyset\}$ must be infinite. For each $m \in M$, define $n(m) = \min\{n \in \mathbb{N} \mid f(n) \in X_m \times \{x\}\}$. Then $f(n(m)) = (x_m, m)$ for a unique element x_m of X_m. Thus $(x_m)_{m\in M} \in \prod_{m\in\mathbb{N}} X_m$. Thus **PCC**(fin) holds. By Exercise E 5 this implies **CC**(fin).

Next, we present some maximality principles:

Definition 2.15. *1.* **PIT**, *the* Boolean Prime Ideal Theorem, *states that every Boolean algebra with* $0 \neq 1$ *has a maximal ideal.*
2. **UFT**, *the* Ultrafilter Theorem, *states that on any set every filter can be enlarged to an ultrafilter.*
3. **UFT**($\mathbb{N}$) *states that on* $\mathbb{N}$ *every filter can be enlarged to an ultrafilter.*
4. **WUF**, *the* Weak Ultrafilter Principle, *states that every infinite set has a free ultrafilter.*
5. **WUF**($\mathbb{N}$) *states that there exists a free ultrafilter on* $\mathbb{N}$.
6. **WUF**(?) *states that there exists a free ultrafilter on some set.*

Theorem 2.16. *1.* **UFT** $\Leftrightarrow$ **PIT**.
2. **UFT** $\Rightarrow$ **AC**(fin).
3. **UFT** $\Rightarrow$ **WUF** $\Rightarrow$ **WUF**($\mathbb{N}$) $\Rightarrow$ **WUF**(?).

Proof. **(1)** See Theorem 4.37.
(2) See (1) and Exercises to Section 4.8, E 9.
(3) Holds trivially.

Next, we present some ordering principles:

Definition 2.17. *1.* **OP**, *the* Ordering Principle, *states that every set can be linearly ordered.*

[14] *Dedekind–infinite* is sometimes abbreviated as *D–infinite*. Likewise *Dedekind–finite* (i.e., not Dedekind–infinite) is sometimes abbreviated as *D–finite*. Cf. Definition 4.1 and Proposition 4.2.

2. **OEP**, *the* Order Extension Principle, *states that every partial order relation on a set can be enlarged to a linear order relation.*

Theorem 2.18. *1.* **UFT** $\Rightarrow$ **OEP**.
2. **OEP** $\Rightarrow$ **OP**.
3. **KW** $\Rightarrow$ **OP**.
4. **OP** $\Rightarrow$ **AC**(fin).

Proof. **(1)** and **(2)**: See Proposition 4.39.
(3) See Proposition 4.40.
(4) Let $(X_i)_{i\in I}$ be a family of non–empty finite sets. Order $X = \bigcup_{i\in I} X_i$ linearly. Then each X_i has a smallest element x_i w.r.t. this order. Thus $(x_i)_{i\in I} \in \prod_{i\in I} X_i$.

Besides conditions weaker than **AC** there are stronger ones. Let us mention the following:

Definition 2.19. *1.* **GCH**, *the* Generalized Continuum Hypothesis, *states that for infinite cardinals a and b the inequalities $a \leq b < 2^a$ imply $a = b$.*
2. **AH**, *the* Aleph–Hypothesis, *states that $2^{\aleph_\alpha} = \aleph_{\alpha+1}$ for each ordinal α.*

The following result, though interesting, we present without proof, since even the special forms of **GCH** resp. **AH**, namely

CH, the *Continuum Hypothesis*: $\aleph_0 \leq b < 2^{\aleph_0} \Rightarrow \aleph_0 = b$,
resp. **AH**(0), the *Special Aleph–Hypothesis*: $\aleph_1 = 2^{\aleph_0}$,

are generally supposed to be false.

Theorem 2.20. *1.* **GCH** $\Rightarrow$ **AC**.
2. **AH** $\Rightarrow$ **AC**.
3. **GCH** $\Leftrightarrow$ **AH**.

Proof. **(1)** See [Sie47], [Spe54].
(2) Immediate from Theorem 2.4, since **AH** implies that the power set of each well–orderable set is well–orderable.
(3) Immediate from (1) and (2), since by Theorem 1.4 each set is well–orderable and thus each cardinal is an Aleph[15].

Observe that **CH** and **AH**(0) are not equivalent. See Section 7.2.

Observe further that the existence of sufficiently many strongly inaccessible cardinals also implies **AC**[16]. However, a discussion of such phenomena is beyond the scope of this book.

---

15 *Alephs* are the cardinals of well–orderable sets.

16 [Tar39]

The following diagram illustrates the logical relations between the principles presented in this section: Note that none of the singleheaded arrows **A** $\rightarrow$ **B**, with the possible exceptions of **CC** $\rightarrow$ **CMC**, **CC**(fin) $\rightarrow$ **CC**(n) and **WUF**($\mathbb{N}$) $\rightarrow$ **WUF**(?) is an equivalence[17].

Diagram 2.21.

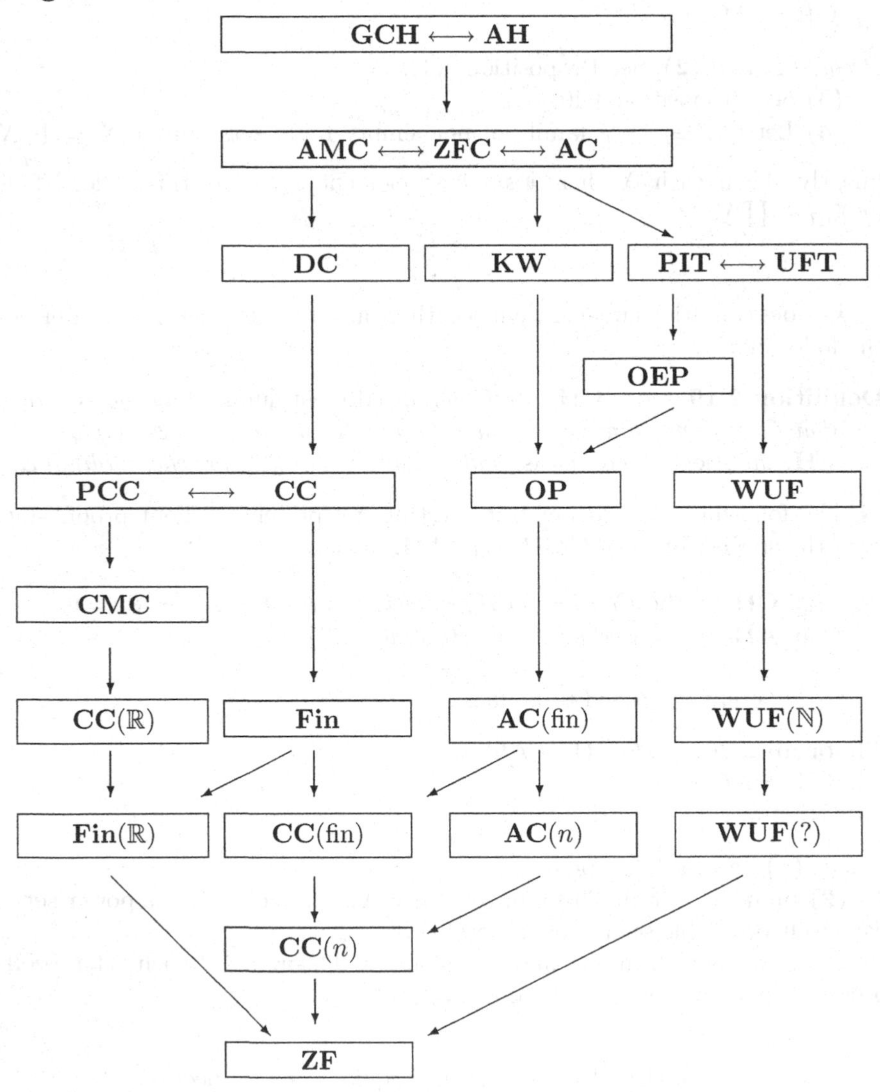

17 [HoRu98]

Exercises to Section 2.2:

E 1. [18] Show that **AC**(2) and **AC**(4) are equivalent.

E 2. Show the equivalence of:
(1) **DC**,
(2) **DMC** (states that for each non–empty set X and each relation ϱ on X, such that for each $x \in X$ there exists some $y \in X$ with $x\varrho y$, there exists a sequence (F_n) of non–empty, finite subsets of X such that for each $n \in \mathbb{N}$ and each $x \in F_n$ there exists some $y \in F_{n+1}$ with $x\varrho y$) and **CC**(fin).

E 3. Show that **DC** is equivalent to the following statement: If ϱ is a relation on a set X such that for each $x \in X$ there exists some $y \in X$ with $x\varrho y$, then for any a in X there exists a sequence (x_n) in X with $x_0 = a$ and $x_n \varrho x_{n+1}$ for each $n \in \mathbb{N}$.

E 4. Show that **CC** is equivalent to the statement: For each sequence $(X_n)_{n\in\mathbb{N}}$ of non–empty sets, there exists a sequence that meets infinitely many X_n's.

E 5. Define **PCC**($\mathbb{R}$), **PCC**(fin), and **PCMC** analogously to **PCC**, and show:
(1) **PCC**($\mathbb{R}$) $\Leftrightarrow$ **CC**($\mathbb{R}$).
(2) **PCC**(fin) $\Leftrightarrow$ **CC**(fin).
(3) **PCMC** $\Leftrightarrow$ **CMC**.
[Hint: Proceed as in proof of Theorem 2.12 (3) using resp. the facts that:
1) Finite products of finite sets are finite.
2) There exists a sequence $(f_n)_{n\in\mathbb{N}^+}$ of bijections $f_n \colon \mathbb{R}^n \to \mathbb{R}$.]

E 6. Show that **CC**($\mathbb{R}$) implies **Fin**($\mathbb{R}$).

E 7. [19] Show that each of the following conditions implies the succeeding ones:
a) **AC**.
b) There are *enough projective sets*, i.e., every set is an image of a projective set (in other words: there exists arbitrary large projective sets w.r.t. the order $\leq^*$ of cardinals).[20]
c) **DC**.

E 8. Show the equivalence of:
a) **CC**.
b) $\mathbb{N}$ is a projective set.

---

[18] [Sie58]
[19] [Blass79]
[20] Cf. with Exercises to Section 2.1, E 4, condition (4). Whether (a) and (b) are equivalent is not known.

E 9. Show that:

a) Every separable pseudometrizable space is second countable.

b) Every second countable pseudometric space is separable iff **CC** holds.

E 10. Show that **AC**($\mathbb{R}$) implies **UFT**($\mathbb{N}$) and thus **WUF**($\mathbb{N}$).
[Hint: By Exercises to Section 1.1, E 1 and $|\mathbb{R}| = |\mathcal{P}\mathbb{N}|$, **AC**($\mathbb{R}$) implies that $\mathcal{P}\mathbb{N}$ is well–orderable.]

3

Elementary Observations

3.1 Hidden Choice

Experience gained presenting the contents of this paper before a learned audience discloses that the wily, attentive listener will expend more energy searching for the possible hidden presence of the axiom of choice in the proofs than he will in following the positive, constructive aspects of these proofs.

W.W. Comfort[1]

It is a historic irony that many of the mathematicians who later opposed the Axiom of Choice had used it implicitly in their own researches.

Gregory H. Moore[2]

The use of the Axiom of Choice is sometimes hidden, and, even if obvious to the expert, may elude the novice. Even several of those mathematicians who rejected **AC** used it unconsciously. Hardy[3] pointed out that Borel, though strongly objecting to the use of **AC** for uncountable indexing sets[4], used it for an indexing set of cardinality $2^{\aleph_0}$ in his proof that there exist continuous functions $f: \mathbb{R} \to \mathbb{R}$ which cannot be represented as double series of polynomials. Sierpiński[5] demonstrated that Lebesgue, another outspoken critic of **AC**, used it to show that countable unions of measurable sets of reals are again measurable. And Moore[6] exhibited a plethora of examples demonstrating that "*future critics of the Axiom* [of Choice] *where freely employing sequences of arbitrary choices in real analysis before Zermelo's proof appeared.*"

---

1 [Com68]
2 [Moo82, p. 64]
3 [Hard06, p. 15]
4 See the headquote for Section 2.2
5 [Sie58, p. 127].
6 [Moo82, §17].

Here follows an instructive example of a proof that appears to be constructive, but is not:

Statement 3.1. Countable unions of countable sets are countable.

Proof. Let $X = \bigcup_{n\in\mathbb{N}} X_n$ be a countable union of countable sets X_n. Assume, without loss of generality, that the X_n's are pairwise disjoint. Since the X_n's are countable they can be written in the form

$$X_n = \{x_n^i \mid i \in \mathbb{N}\} = \{x_n^0, x_n^1, x_n^2, \ldots\}.$$

Define a bijection $f\colon \mathbb{N} \to X$ via the following construction:

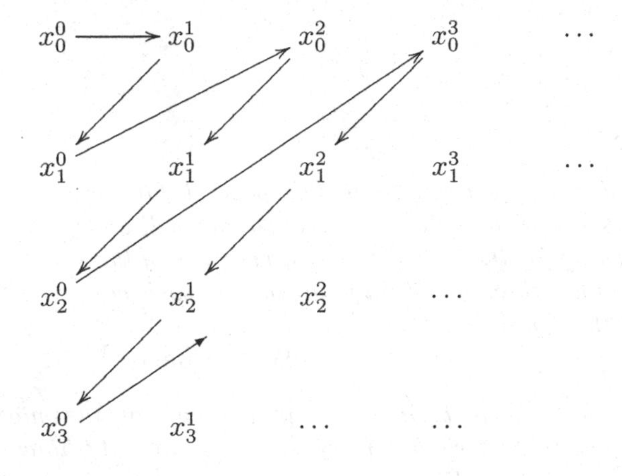

i.e.,

$$f(0) = x_0^0, \quad f(1) = (x_0^1), \quad f(2) = x_1^0, \quad f(3) = (x_0^2), \quad f(4) = x_1^1,$$
$$f(5) = (x_2^0), \quad f(6) = x_0^3, \quad \cdots$$

resp. $g = f^{-1}\colon X \to \mathbb{N}$ by the explicit formula:

$$g(x_i^k) = \sum_{\nu=1}^{i+k} \nu + (i+1) = \frac{(i+k)\cdot(i+k+1)}{2} + (i+1).$$

Thus X is countable.

Discussion: Once the X_n's are expressed in the form $X_n = \{x_n^i \mid i \in \mathbb{N}\}$, the remainder of the above proof is constructive indeed. However, there are many ways to express each X_n in the above form, i.e., as a *counted* set instead of a *countable* set. Thus the above proof uses **CC**, the Axiom of Countable Choice.

Let us analyze the situation in more detail by introducing some definitions first:

Definition 3.2. *1.* **CUT**, *the* Countable Union Theorem, *states that countable unions of at most countable sets are at most countable.*

2. **CUT**($\mathbb{R}$) *states that countable unions of at most countable subsets of* $\mathbb{R}$ *are at most countable.*

3. **CUT**(fin) *states that countable unions of finite sets are at most countable.*

4. **CUT**(2) *states that countable unions of 2–element sets are at most countable.*

Then the proof given for Statement 3.1 yields:

Proposition 3.3. CC *implies* **CUT**.

The next diagram illustrates further relations between the above and some closely related concepts: note that none of the single–headed arrows $\mathbf{A} \to \mathbf{B}$, with the possible exception of "*Lebesgue measure is countably additive* $\to$ $\mathbb{R}$ *is not a countable union of countable sets*" is an equivalence.

Note further that none of the conditions entering the diagram holds in **ZF**. In particular there exist models of **ZF** in which $\mathbb{R}$ is a countable union of countable sets[7], and models of **ZF** in which a countable union of certain 2–element sets is uncountable[8].

The proofs of the two equivalences will follow after the diagram.

Diagram 3.4.

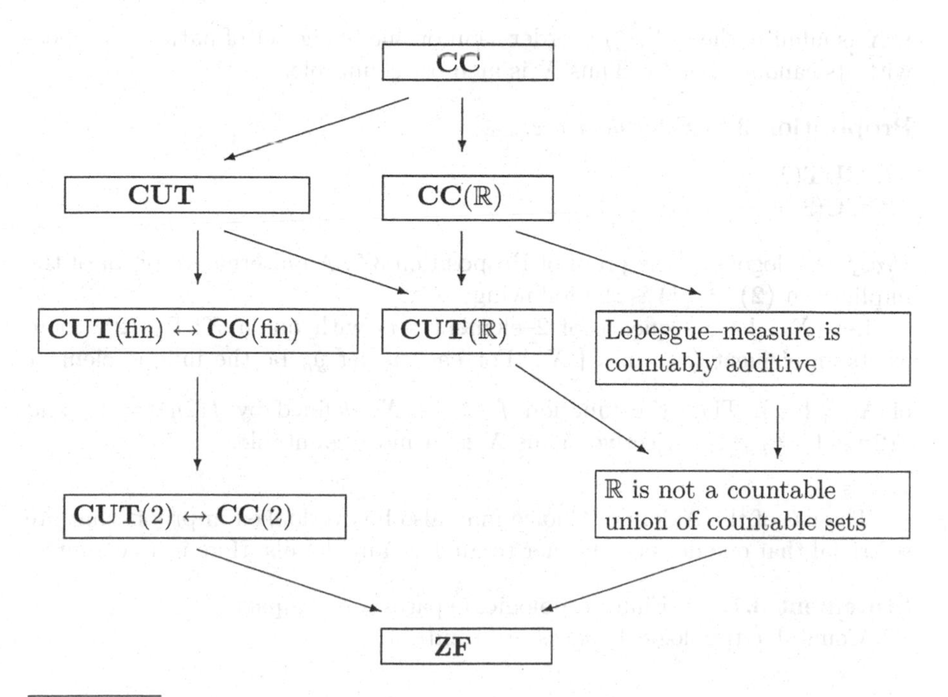

[7] e.g., in the Feferman–Levy Model A8 (M9 in [HoRu98]).

[8] e.g., in Cohen's second model (M7 in [HoRu98]).

Proposition 3.5. *Equivalent are:*

1. **CUT**(fin).
2. **CC**(fin).

Proof. **(1) ⇒ (2)** Let (X_n) be a sequence of non–empty finite sets. By (1), there exists a surjection $f\colon \mathbb{N} \to \bigcup\limits_{n\in\mathbb{N}} X_n$. For each n, define

$$x_n = f(\min\{m \in \mathbb{N} \mid f(m) \in X_n\}).$$

Then $(x_n) \in \prod\limits_{n\in\mathbb{N}} X_n$.

(2) ⇒ (1) Let (X_n) be a sequence of finite sets with $\bigcup\limits_{n\in\mathbb{N}} X_n = X$. Then the sequence (Y_n), defined by $Y_n = X_n \setminus \bigcup\limits_{m<n} X_m$ is a sequence of pairwise disjoint finite sets with $\bigcup\limits_{n\in\mathbb{N}} Y_n = X$. For each n, the set L_n of linear order relations on Y_n is a non–empty, finite set. By (2), there exists an element (φ_n) in $\prod\limits_{n\in\mathbb{N}} L_n$. Order X as follows

$$x \leq y \Leftrightarrow \begin{cases} x \in Y_n \text{ and } y \in Y_m \text{ and } n < m \\ \text{or} \\ \{x, y\} \subseteq Y_n \text{ and } x\varphi_n y. \end{cases}$$

If X is infinite, then $(X, \leq)$ is order–isomorphic to the set of natural numbers with its canonical order. Thus X is at most countable.

Proposition 3.6. *Equivalent are:*

1. **CUT**(2).
2. **CC**(2).

Proof. Analogous to the proof of Proposition 3.5. An alternative proof of the implication **(2) ⇒ (1)** is the following:

Let (X_n) be a sequence of 2–element sets with union X: By (2), there exists an element (x_n) in $\prod\limits_{n\in\mathbb{N}} X_n$. For each n, let y_n be the unique element of $X_n \setminus \{x_n\}$. Then the function $f\colon \mathbb{N} \to X$, defined by $f(2n) = x_n$ and $f(2n+1) = y_n$, is surjective. Thus X is at most countable.

The use of the Axiom of Choice may also be overlooked in proofs that are so trivial that one does not bother to analyze any details. Here is an example:

Statement 3.7. 1. Finite topological spaces are compact.
2. Countable topological spaces are Lindelöf.

Proof. Obvious.

Discussion: Although the proofs of the statements (1) and (2) are straightforward indeed, and very similar too, there is a marked difference. The proof of (1) holds in **ZF**, the proof of (2), however, makes use of the Axiom of Choice. So (1) is a result in **ZF**, but (2) fails to be so. In fact, the following holds:

Theorem 3.8. [9] *Equivalent are:*

1. *Countable topological spaces are Lindelöf.*
2. *$\mathbb{N}$ is Lindelöf.*
3. *$\mathbb{Q}$ is Lindelöf.*
4. *Every unbounded subset of $\mathbb{R}$ contains an unbounded sequence.*
5. **CC**$(\mathbb{R})$.

Proof. **(1) ⇒ (3)** since $\mathbb{Q}$ is countable.

(3) ⇒ (4). Let A be a subset of $\mathbb{R}$, unbounded to the right. Let, for each $a \in A$, $C(a) = \{x \in \mathbb{Q} \mid x < a\}$. Then $\mathcal{C} = \{C(a) \mid a \in A\}$ is an open cover of $\mathbb{Q}$. Let $\mathcal{C}' = \{C_n \mid n \in \mathbb{N}\}$ be a countable subset of $\mathcal{C}$ that covers $\mathbb{Q}$. For each C_n there exists a unique element a_n in A with $C_n = C(a_n)$ — since $\sup_{\mathbb{R}} C(a) = a$. Thus (a_n) is an unbounded sequence in A.

(4) ⇒ (5) Let (X_n) be a sequence of non-empty subsets X_n of $\mathbb{R}$. Let $f\colon \mathbb{R} \to (0,1)$ be a fixed bijection, and, for each $n \in \mathbb{N}$, let $\sigma_n\colon \mathbb{R} \to \mathbb{R}$ be defined by $\sigma_n(x) = n + x$.

Define $Y_n = \sigma[f[X_n]]$. Then $Y_n \subseteq (n, n+1)$, and $Y = \bigcup_{n\in\mathbb{N}} Y_n$ is an unbounded subset of $\mathbb{R}$.

By (4), there exists an unbounded sequence y_n in Y. Thus $M = \{m \in \mathbb{N} \mid \exists n \in \mathbb{N}\ \ y_n \in Y_m\}$ is an infinite subset of $\mathbb{N}$ with $\prod_{m\in M} Y_m \neq \emptyset$. Let $(\tilde{y}_m)_{m\in M}$ be an element of this product. Define x_m to be the unique element of X_m with $\sigma_m(f(x_m)) = \tilde{y}_m$. Then $(x_m) \in \prod_{m\in M} X_m$. Thus (5) follows via Exercises to Section 2.2, E 5(1).

(5) ⇒ (2) Let $\mathfrak{U}$ be an open cover of $\mathbb{N}$. Then for each $n \in \mathbb{N}$ the set $X_n = \{U \in \mathfrak{U} \mid n \in U\}$ is a non–empty subset of $\mathcal{P}\mathbb{N}$, the powerset of $\mathbb{N}$. Since $\mathcal{P}\mathbb{N}$ and $\mathbb{R}$ have the same cardinality, (5) implies that there exists an element (U_n) in $\prod_{n\in\mathbb{N}} X_n$. Thus $\{U_n \mid n \in \mathbb{N}\}$ is a countable subcover of $\mathfrak{U}$.

(2) ⇒ (1) Every countable topological space is a continuous image of $\mathbb{N}$. Since continuous images of Lindelöf spaces are Lindelöf, (1) follows from (2).

In analogy to Statement 3.7 we obtain:

Statement 3.9. 1. Finite sums of compact topological spaces are compact.
2. Countable sums of Lindelöf spaces are Lindelöf.

[9] [Bru82a], [HeSt97].

Proof. Obvious.

Discussion: Here, as for Statement 3.7, (1) and (2) have trivial and parallel proofs, but while (1) is a result in **ZF**, (2) is not. In fact, as will be shown in Theorem 4.62, (2) is equivalent to **CC**. Not even finite sums of Lindelöf spaces need to be Lindelöf. Even worse, the sum of a compact space and a Lindelöf space may fail to be Lindelöf, as shown in Remark 4.63.

Exercises to Section 3.1:

E 1. Show that for each natural number n the following conditions are equivalent:
- (1) **CC**($\leq n$), i.e., $\prod_{i\in\mathbb{N}} X_i \neq \emptyset$ for each sequence $(X_i)_{i\in\mathbb{N}}$ of sets with $1 \leq |X_i| \leq n$.
- (2) For each sequence $(X_i)_{i\in\mathbb{N}}$ of n–element sets there exists a sequence $(\leq_i)_{i\in\mathbb{N}}$ of linear orders $\leq_i$ on X_i.
- (3) **CUT**(n), i.e., countable unions of n–element sets are at most countable.
- (4) **CUT**($\leq n$), i.e., the union of each sequence $(X_i)_{i\in\mathbb{N}}$ of sets with $1 \leq |X_i| \leq n$ is at most countable.
- (5) **CC**(i) holds for each $i \in \{1, 2, \ldots, n\}$.

E 2. Show the equivalence of the following conditions:
- (1) **PCC**(2), i.e., for each sequence (X_n) of 2–element sets X_n there exists an infinite subset M of $\mathbb{N}$ with $\prod_{m\in M} X_m \neq \emptyset$.
- (2) The countable union of pairwise disjoint 2–element sets is Dedekind–infinite.

E 3. Show that, for each at most countable set X, the set $\bigcup_{n\in\mathbb{N}} X^n$ is at most countable.

E 4. Show that the following sets are countable:
- (1) The set $\mathbb{Q}$ of rational numbers.
- (2) The set $\bigcup_{n\in\mathbb{N}} \mathbb{N}^n$.
- (3) The set of all algebraic real numbers.
- (4) The set of all algebraic complex numbers.

E 5. [10] Show that the following condition (1) implies all subsequent ones:
- (1) $\mathbb{R}$ is the countable union of countable sets.
- (2) ω_1 is not regular, i.e., there exists a sequence $(\alpha_n)_{n\in\mathbb{N}}$ of countable ordinals with $\omega_1 = \sup_{n\in\mathbb{N}} \alpha_n$.
- (3) **Fin**($\mathbb{R}$).

[10] [Spe57]

(4) $\aleph_1$ and $2^{\aleph_0}$ are incomparable w.r.t. $\leq$.
Moreover[11], show that (2) implies that **CC**($\mathbb{R}$) fails.
[Hint: For (1) $\Rightarrow$ (2) use Exercise to Section 4.2, E 4.].

E 6. [12] Show that **CC**($\mathbb{R}$) implies that the Lebesgue–measure is σ-additive.
[Hint: See the proof of Proposition 7.14. Cf. also Exercise to Section 5.1, E 13.]

3.2 Unnecessary Choice

In Section 3.1 it has been shown that several **ZFC**–theorems may erroneously be considered to hold in **ZF**; due to the fact that the use of **AC** was hidden. Here we will show that several **ZFC**–theorems may erroneously be considered to fail in **ZF**; due to the fact that their familiar proofs use **AC**, although this use of **AC** may be only apparent or can be avoided by alternative proofs which sometimes are drastically different from the familiar ones but more often just require a minor adjustment of the latter.

Proposition 3.10. *Finite sums of compact spaces are compact.*

Proof in **ZFC**: Let X be the sum of the pairwise disjoint compact spaces $X_1, \ldots, X_n$, and let $\mathcal{B}$ be an open cover of X. For each $i = 1, \ldots, n$ the set $\mathcal{B}_i = \{B \cap X_i \mid B \in \mathcal{B}\}$ is an open cover of X_i. By compactness each $\mathcal{B}_i$ contains a finite cover $\mathcal{F}_i$ of X_i. For each F in $\mathcal{F}_i$ choose an element $B(F)$ in $\mathcal{B}$ with $F = B(F) \cap X_i$.
Then the set $\mathcal{F} = \{B(F) \mid i \in I \text{ and } F \in \mathcal{F}_i\}$ is a finite cover of X with $\mathcal{F} \subseteq \mathcal{B}$. Thus X is compact.

Observation: In the above proof, choice has been used to select the $B(F)$'s. However, this use of choice is only apparent, since I and each $\mathcal{F}_i$ are finite, and the axiom of choice for finite indexing sets holds in **ZF**. Thus the above is a valid proof in **ZF**. Observe however that the analogous proof of the statement

Finite sums of Lindelöf spaces are Lindelöf

fails in **ZF**, since here countable choice is used. Indeed, the statement about Lindelöf spaces may fail in **ZF**. See Remark 4.63.

Theorem 3.11. *Every bounded, infinite subset X of $\mathbb{R}$ has an accumulation point*[13] *in $\mathbb{R}$.*

11 [Chu27]
12 [KeTa2003]
13 x is called an *accumulation point* of X iff every neighborhood of x meets X in an infinite set.

Proof in **ZFC**: Since X is bounded, there exist real numbers a and b with $X \subseteq [a, b]$. Observe first that $[a, b]$ is compact. Proceed indirectly by assuming that there exists an open cover $\mathcal{B}$ of $[a, b]$ without finite subcover, define $x = \sup\{y \in [a, b] \mid \text{some finite subset of } \mathcal{B} \text{ covers } [a, y]\}$, observe that $x \in B$ for some $B \in \mathcal{B}$, and arrive at a contradiction. Next, assume that X has no accumulation point in $\mathbb{R}$. Then for each $x \in [a, b]$, there exists an open neighborhood $B(x)$ such that $X \cap B(x)$ is finite. Then $\mathcal{B} = \{B(x) \cap [a, b] \mid x \in [a, b]\}$ is an open cover of $[a, b]$, and thus has a finite subcover $\mathcal{F}$. Consequently $X = \bigcup_{F \in \mathcal{F}} (X \cap F)$ is a finite union of finite sets and thus finite, a contradiction.

Observation: In the above proof, choice has been used to select the $B(x)$'s. This use of choice can be avoided by defining $\mathcal{B}$ alternatively as the set of all open subsets of $[a, b]$ that meet only finitely many elements of X. This way a proof in **ZF** results.

The method, employed above, to modify a **ZFC**–proof into a **ZF**–proof, can be applied quite often. Here is another example:

Proposition 3.12. *Closed subspaces of compact spaces are compact.*

Proof in **ZFC**: Let X be a closed subspace of a compact space Y, and let $\mathcal{B}$ be an open cover of X. For each $B \in \mathcal{B}$ select an open set $A(B)$ in Y with $B = X \cap A(B)$. Define $\mathcal{A} = \{A(B) \mid B \in \mathcal{B}\}$. Then $\mathcal{A} \cup \{Y \setminus X\}$ being an open cover of Y, contains a finite cover $\mathcal{F}$. Consequently $\mathcal{G} = \{X \cap F \mid F \in (\mathcal{F} \cap \mathcal{A})\}$ is a finite cover of X with $\mathcal{G} \subseteq \mathcal{B}$, Thus X is compact.

Observation: In the above proof, choice has been used to select the $A(B)$'s. This use of choice can be avoided by defining $\mathcal{A}$ alternatively by

$$\mathcal{A} = \{A \mid A \text{ open in } Y \text{ and } (X \cap A) \in \mathcal{B}\}.$$

Theorem 3.13. [14] *$[0, 1]^{\mathbb{N}}$ is compact.*

Proof in **ZFC**: It suffices (cf. the remark following Definition 3.21) to show that in $X = [0, 1]^{\mathbb{N}}$ every filter $\mathcal{F}$ has a cluster point (x_n). Denote the n–th projection by $\pi_n\colon [0, 1]^{\mathbb{N}} \to [0, 1]$, and the neighborhood–filter of a point x in $[0, 1]$ by $\mathcal{U}(x)$. Define, by recursion, points x_n in $[0, 1]$ and filters $\mathcal{F}_n$ on X as follows:

x_0 is a cluster point of the filter $\{G \subseteq [0, 1] \mid \pi_0^{-1}[G] \in \mathcal{F}\}$.
$\mathcal{F}_0$ is the filter on X, generated by the set

$$\mathcal{F} \cup \{\pi_0^{-1}[U] \mid U \in \mathcal{U}(x_0)\}.$$

x_{n+1} is a cluster point of the filter $\{G \subseteq [0, 1] \mid \pi_{n+1}^{-1}[G] \in \mathcal{F}_n\}$.
$\mathcal{F}_{n+1}$ is the filter on X, generated by the set

$$\mathcal{F}_n \cup \{\pi_{n+1}^{-1}[U] \mid U \in \mathcal{U}(x_{n+1})\}.$$

---

[14] [Loe65], [DHHKR2003a].

Then $x = (x_n)$ is a cluster point of $\bigcup_{n\in\mathbb{N}} \mathcal{F}_n$ and thus of $\mathcal{F}$ in X.

Observation: In the above proof, choice has been used (in fact: dependent choice) to select the x_n's. This use of choice can be avoided via the observation that for every filter $\mathcal{G}$ on $[0,1]$ the set of cluster points of $\mathcal{G}$, being a non–empty, closed subset of $[0,1]$, contains a smallest member. By choosing the x_n's as the smallest cluster points of the corresponding filters we obtain a proof in **ZF**.

Another such example, using distinguished choices instead of arbitrary ones, is the following:

Proposition 3.14. *Every continuous function $f\colon [0,1] \to \mathbb{R}$ is uniformly continuous.*

Proof in **ZFC**: Let ϵ be a positive real number. For every $x \in [0,1]$, select a positive δ_x such that $y \in [[0,1]\cap(x-\delta_x, x+\delta_x)]$ implies $|f(y)-f(x)| \leq \frac{\epsilon}{2}$. Then the sets $U_x = [0,1]\cap(x-\delta_x, x+\delta_x)$ form, for $x \in [0,1]$, an open cover of $[0,1]$. By compactness, there exists a finite subset F of $[0,1]$ such that $\{U_x \mid x \in F\}$ covers $[0,1]$. Consider $\delta = \min\{\frac{\delta_x}{2} \mid x \in F\}$. Then, for any points x and y in $[0,1]$ with $|x-y| < \delta$, there exists some $z \in F$ with $\{x, y\} \subseteq U_z$, which implies $|f(x) - f(y)| \leq |f(x) - f(z)| + |f(z) - f(y)| < \epsilon$.

Observation: In the above proof, choice has been used to select the δ_x's. However, as in the previous result, choice can be avoided, e.g., by defining $\delta_x = \frac{1}{n_x}$, where $n_x = \min\{n \in \mathbb{N}^+ \mid y \in [[0,1]\cap(x-\frac{1}{n}, x+\frac{1}{n})] \Rightarrow |f(y)-f(x)| < \frac{\epsilon}{2}\}$.

Next, an example of a **ZFC**–proof which requires drastic remodeling.

Theorem 3.15. [15] *Every sequentially continuous*[16] *function $f\colon \mathbb{R} \to \mathbb{R}$ is continuous.*

Proof in **ZFC**: Assume that f is sequentially continuous but not continuous at some $x \in \mathbb{R}$. Then there exists a positive real number ϵ such that for each positive real number δ there exists some $y \in \mathbb{R}$ with $|x - y| \leq \delta$ and $|f(x) - f(y)| > \epsilon$. Choose, for every $n \in \mathbb{N}^+$, an x_n in $\mathbb{R}$ with

$$|x - x_n| \leq \frac{1}{n} \text{ and } |f(x) - f(x_n)| > \epsilon.$$

Then the sequence (x_n) converges to x, but the sequence $(f(x_n))$ does not converge to $f(x)$, contradicting the sequential continuity of f at x.

---

15 [Sie18]

16 f is called *sequentially continuous* iff $(x_n) \to x$ implies $(f(x_n)) \to f(x)$.

Observation: In the above proof, choice has been used to select the x_n's. This use of choice is essential. In fact, the statement

> *Every function $f\colon \mathbb{R} \to \mathbb{R}$, that is sequentially continuous at some point x, must be continuous at x*

does not hold in **ZF** (see Theorem 4.54). However, Theorem 3.15 holds in **ZF**:

Proof in **ZF**: Let $f\colon \mathbb{R} \to \mathbb{R}$ be sequentially continuous. Using the fact that $\mathbb{Q}$ is countable and is dense in $\mathbb{R}$ we obtain, as above, that for each $x \in \mathbb{R}$ the restriction of f to the set $\mathbb{Q} \cup \{x\}$ is continuous at x. Thus for $\epsilon > 0$ there exists $\delta > 0$ such that

$$(y \in \mathbb{Q} \text{ and } |x - y| \leq \delta) \text{ implies } |f(x) - f(y)| \leq \epsilon.$$

Since for each $z \in \mathbb{R}$ with $|x - z| \leq \delta$, the restriction of f to $\mathbb{Q} \cup \{z\}$ is continuous, the above implies that $|f(x) - f(z)| \leq \epsilon$. Thus f is continuous at x.

Finally, two examples of **ZFC**–proofs that remain valid only under special assumptions. If these are not satisfied, another proof has to be constructed. In other words: these results require two distinct proofs, each covering just some part of the **ZF**–world:

Theorem 3.16. [17] *For subspaces X of $\mathbb{R}$, the following conditions are equivalent:*

1. *X is compact.*
2. *X is sequentially compact and Lindelöf.*

Proof in **ZFC**: Obviously (1) implies (2). For the converse, assume that there exists an open cover $\mathcal{B}$ of X without a finite subcover. Since X is Lindelöf, we may assume that $\mathcal{B}$ is countable, say $\mathcal{B} = \{B_n \mid n \in \mathbb{N}\}$. We may further assume that, for each $n \in \mathbb{N}$, $B_n \subsetneqq B_{n+1}$ and $B_{n+1} \setminus B_n \neq \emptyset$. Choose elements x_n in $B_{n+1} \setminus B_n$. Then (x_n) is a sequence in X without a convergent subsequence, contradicting sequential compactness of X. Thus X is compact.

Observation: In the above proof, choice has been used to select the x_n's. Since the sets $B_{n+1} \setminus B_n$ are subsets of $\mathbb{R}$, the proof remains valid provided **CC**($\mathbb{R}$) is satisfied. What happens if **CC**($\mathbb{R}$) fails? Here a completely different proof is needed. This will be supplied by Theorem 7.2, where it is shown that the failure of **CC**($\mathbb{R}$) implies that every Lindelöf–subspace of $\mathbb{R}$ is already compact.

Theorem 3.17. [18] *Countable products of 2–element topological spaces are compact.*

---

[17] [Gut2003]
[18] [HeKe2000]

Proof in **ZFC**: Let $(X_n)_{n\in\mathbb{N}}$ be a sequence of topological spaces X_n, each having precisely two points x_n and y_n. Let $\mathbf{2}$ be the discrete space with underlying set $\{0,1\}$. Then for each $n \in \mathbb{N}$, the map $f_n\colon \mathbf{2} \to X_n$, defined by $f_n(0) = x_n$ and $f_n(1) = y_n$ is a continuous surjection. Hence $\prod_{n\in\mathbb{N}} X_n$ is a continuous image of $\mathbf{2}^{\mathbb{N}}$. Since the latter, being a closed subspace of $[0,1]^{\mathbb{N}}$, is compact by Proposition 3.12 and Theorem 3.13, so is the former.

Observation: In the above proof choice has been used to define the f_n's, since we have no rule that determines which of the 2 points of X_n is to be called x_n, and which y_n. However, the following simple case–distinction provides a **ZF**–proof (each of the 2 cases may indeed occur):

Case 1: $\prod_{n\in\mathbb{N}} X_n = \emptyset$. In this case $\prod_{n\in\mathbb{N}} X_n$ is trivially compact.

Case 2: $\prod_{n\in\mathbb{N}} X_n \neq \emptyset$. In this case, let (x_n) be an element of $\prod_{n\in\mathbb{N}} X_n$. Then proceed as in the **ZFC**–proof, described above.

Observe, however, that in **ZF** countable products of 3–element spaces may fail to be compact. See Section 4.8 for more details.

Exercises to Section 3.2:

E 1. Show that every closed subspace of a Lindelöf space is Lindelöf.

E 2. Show that, in a topological space, a set is open iff it is a neighborhood of each of its points.

E 3. Show that a topological space X is compact iff each open cover $\mathcal{B}$ of X contains a finite refinement $\mathcal{F}$ (i.e., a finite set $\mathcal{F}$ such that $\bigcup \mathcal{F} = X$ and for each $F \in \mathcal{F}$ there is a $B \in \mathcal{B}$ with $F \subseteq B$).

E 4. Show that:
(a) Every compact Hausdorff space is normal.
(b) Every regular Lindelöf space is normal.

E 5. Let A be a dense, D–finite subset of $\mathbb{R}$. (Such sets exist, provided that infinite, D–finite subsets of $\mathbb{R}$ exist[19]). Define $f\colon \mathbb{R} \to \mathbb{R}$ by $f(x) = \begin{cases} 1, \text{ if } x \in A \\ 0, \text{ otherwise.} \end{cases}$
Show that:
(a) f is sequentially continuous at x iff $x \in (\mathbb{R} \setminus A)$.
(b) f is continuous at no point of $\mathbb{R}$.

E 6. [20] Show that $[0,1]^I$ is compact for each well–orderable set I.

[19] [Bru82]

[20] However $[0,1]^{\mathbb{R}}$ may fail to be compact. See [Ker2000] and Exercises to Section 7.2, E 8. Cf. Exercises to Section 4.8, E 13.

3.3 Concepts Split Up: Compactness

It is a hopeless endeavour, doomed to failure to attempt to prove either the Stone–Čech compactification theorem or the Tychonoff product theorem without invoking some form of the axiom of choice, ... It is my feeling, however, that the definition of compactness relative to which the theorems of Stone–Čech and Tychonoff are unprovable without the axiom of choice is, from the point of view of topological analysis and the theory of rings of continuous functions, unnatural and unsuitable.

W.W. Comfort[21]

However, this is 'mere matter of detail' as the Irishman said when he was asked **how** *he had killed his landlord.*

Thomas Henry Huxley[22]

Another problem, encountered when working in **ZF** instead of **ZFC**, arises from the fact that various familiar descriptions of a certain concept, equivalent to each other in the presence of **AC**, may separate in its absence into different concepts and that the validity in **ZF** of familiar results concerning this concept may depend crucially on the chosen variant of the concept.

Paradigmatically we will illustrate this situation by analyzing the *compactness* concept. In this section we will present several familiar descriptions of compactness and investigate the set–theoretical conditions responsible for any pair of these descriptions to characterize the same concept. How certain theorems, in particular the Tychonoff Theorem and the Čech–Stone Theorem, the Ascoli Theorem, and the Baire Category Theorem, depend on the chosen form will be analyzed in later sections.

Let us start by presenting the definition of compactness (originally termed *bicompactness*) as given by Alexandroff and Urysohn in their fundamental paper[23] (in the original French). They start with the definition of *complete accumulation points*:

Définition 3.18. *Un point est dit point d'accumulation complète de l'ensemble A, si quel-que soi le voisinage $V(\zeta)$ la puissance de $A \cap V(\zeta)$ est égale à celle de A tout entier.*

Next they prove in the following theorem the equivalence of 3 conditions:

Théorème 3.19. I Les trois propriétés suivantes d'un espace topologique R, sont équivalents:

---

21 [Com68]

22 From a letter of T.H. Huxley to his son, April 21, 1879. Quoted from L. Huxley *Life and Letters of Thomas Henry Huxley.* Vol. II, p. 8 (1901).

23 [AlUr29]

(A) Tout ensemble infini situé dans R possède au moins un point d'accumulation complète.
(B) Toute suite infinie bien ordonnée d'ensembles fermés décroissants posséde au moins un point appartenant à tous les ensembles de la suite.
(C) De toute infinité de domaines recouvrant l'espace R, on peut extraire un nombre fini de domaines jouissant de la même propriété.

Finally, after proving the above theorem, they call a space *(bi)compact* provided it satisfies one and hence all three of the above conditions:

Le théorème I justifie, il nous semble la définition fondamentale suivante:

Définition 3.20. *Un espace R s'appelle BICOMPACT s'il vérifie l'un quelconque et par suite toutes les trois conditions* $(A), (B), (C)$.

This procedure of defining mathematical concepts is not uncommon. Unfortunately in the present case Theorem 3.19 fails badly in **ZF**. No two of the 3 properties $(A), (B)$ and (C) remain equivalent. Moreover, as we will see in a moment, the equivalence of conditions (A) and (C) holds true if and *only* if **AC** is valid. Furthermore, in the absence of **AC** condition (B) is somewhat unnatural and impracticable.

So, what should we understand by *compactness*, when working in **ZF**? Historically, condition (A) has been used by early investigators, e.g., Tychonoff and Čech. Most modern books, however, use condition (C). In addition other useful descriptions, in particular by means of filters and ultrafilters, have emerged over the years. Next we are going to investigate the relations between the most familiar compactness concepts; first in the realm of topological spaces, next in that of completely regular spaces, then for pseudometric spaces, and finally for subspaces of $\mathbb{R}$. Interesting facts will emerge on each of these levels.

Definition 3.21. [24] *A topological space X is called:*

1. Compact *provided that in X every open cover contains a finite one.*
2. Filter–compact *provided that in X each filter has a cluster point.*
3. Ultrafilter–compact *provided that in X each ultrafilter converges.*
4. Alexandroff–Urysohn–compact *provided that in X each infinite subset has a complete accumulation point.*

The only implications between the 4 compactness concepts, defined above, are the trivial ones:

$$\text{compact} \iff \text{filter–compact} \implies \text{ultrafilter–compact}.$$

A compact space need not be Alexandroff–Urysohn–compact (see the proof of Theorem 3.22). Even the closed unit interval $[0, 1]$, which is clearly compact,

[24] For further compactness versions see [Com68], [BeHe98], and [DHHRS2002].

may fail to be Alexandroff–Urysohn–compact (see Theorem 3.32). Conversely, in certain models of **ZF** there exist Alexandroff–Urysohn–compact spaces that fail to be ultrafilter–compact.[25]

Theorem 3.22. [26]

1. *Equivalent are:*
   a) *Compact = Ultrafilter–compact.*
   b) **UFT**, *the Ultrafilter Theorem.*
2. *Equivalent are:*
   a) *Compact = Alexandroff–Urysohn–compact.*
   b) *Ultrafilter–compact = Alexandroff–Urysohn–compact.*
   c) **AC**.

Proof. **(1) (a) $\Rightarrow$ (b)**. By Theorem 4.37, it suffices to show that products of compact Hausdorff spaces are compact. Since products of ultrafilter–compact Hausdorff spaces are ultrafilter–compact, this follows from (a).

(b) $\Rightarrow$ (a) It suffices to show that each filter $\mathcal{F}$ in an ultrafilter–compact space X has a cluster point. By (b), $\mathcal{F}$ can be enlarged to an ultrafilter $\mathcal{U}$. Since every convergence point of $\mathcal{U}$ is a cluster point of $\mathcal{F}$, the result follows.

(2) As is well–known, (c) implies (a) and (b). To show that each of (a) resp. (b) imply (c), consider two infinite cardinals a and b. Then there exist disjoint sets A and B with $|A| = a$ and $|B| = b$. The topological space X whose underlying set is $A \cup B$ and whose open sets are $\emptyset, A, B$, and $A \cup B$, is simultaneously compact and ultrafilter–compact. So each of (a) and (b) imply that X is Alexandroff–Urysohn–compact. Consequently the set $A \cup B$ has a complete accumulation point x. If $x \in A$, then $a = |A| = |A \cup B| \geq b$. If $x \in B$, then $b = |B| = |A \cup B| \geq a$. Thus $a \leq b$ or $b \leq a$. Consequently Theorem 4.20, together with the observation that every finite cardinal is comparable with any other one, implies that **AC** holds.

More important than compact spaces are compact Hausdorff spaces. This is mainly due to two facts. First, compact Hausdorff spaces are regular and satisfy a property that is analoguous to completeness among metric spaces: they are *H–closed*, i.e., closed in every Hausdorff space in which they can be embedded; in other words: they cannot be densely embedded into any properly larger Hausdorff space. In fact, as Alexandroff and Urysohn have shown, these properties characterize compact Hausdorff spaces: they are precisely the H–closed regular spaces. This characterization remains valid in **ZF** (see Exercise E 2).

Secondly, and even more important, is the fact that compact Hausdorff spaces are normal and hence completely regular, — thus, modulo homeomorphism, precisely the closed subspaces of powers $[0,1]^I$ of the closed unit

---

[25] [DHHRS2002]

[26] [How90], [Her96].

interval. Unfortunately, in **ZF** this characterization breaks down for two reasons: On one hand, though compact Hausdorff spaces are still normal, they may no longer be completely regular[27]. On the other hand, though spaces of the form $[0,1]$, $[0,1]^n$, and $[0,1]^{\mathbb{N}}$ are compact[28]; for arbitrary indexing sets I, the spaces $[0,1]^I$ need no longer be so.

Definition 3.23. *A topological space is called* Tychonoff–compact *provided it is homeomorphic to a closed subspace of some power* $[0,1]^I$ *of the closed unit interval.*

Proposition 3.24. [29] *Equivalent are:*

1. *Tychonoff–compact = compact and completely regular.*
2. **UFT**.

Proof. **(1) ⇒ (2)** Theorem 4.70, and the fact that all spaces of the form $[0,1]^I$ are compact implies (2).

(2) ⇒ (1) Every completely regular space X can be embedded into $[0,1]^{C(X,[0,1])}$. If X is compact, H–closedness of X implies that the embedding is closed. Thus X is Tychonoff–compact. Conversely, let X be Tychonoff–compact. Then X is completely regular. By (2) and Theorem 4.70, together with the fact that closed subspaces of compact spaces are compact, X must be compact.

Next we proceed to the realm of pseudometric[30] spaces. Here, as opposed to the situation with metric spaces, where the relations between the various compactness concepts are not yet completely understood, we will be able to satisfactorily analyze the relations between various familiar descriptions of compactness.

Definition 3.25. *A pseudometric space X is called:*

1. Weierstrass–compact *provided in X every infinite set has an accumulation point*[31].
2. Countably compact *provided in X every countable open cover contains a finite one.*
3. Sequentially compact *provided in X every sequence has a convergent subsequence.*
4. Complete *provided in X every Cauchy sequence converges.*

---

[27] [Laeu62/63]

[28] [Loe65]. Cf. Theorem 3.13.

[29] [Her96]

[30] Here $d(x,y)=0$ is allowed for distinct points x and y.

[31] x is called an *accumulation point* of A iff every neighborhood of x meets A in an infinite set.

5. Totally bounded *provided that for each positive real number r there exists a finite subset F of X such that for each $x \in X$ there exists some $y \in F$ with $d(x, y) < r$.*

Proposition 3.26. [32] *Under* **CC** *the following conditions are equivalent for pseudometric spaces X:*

1. *X is compact.*
2. *X is Weierstrass–compact.*
3. *X is countably compact.*
4. *X is sequentially compact.*
5. *X is totally bounded and complete.*

Proof. The implications, illustrated in the following diagram, are easily shown to hold even in **ZF**:

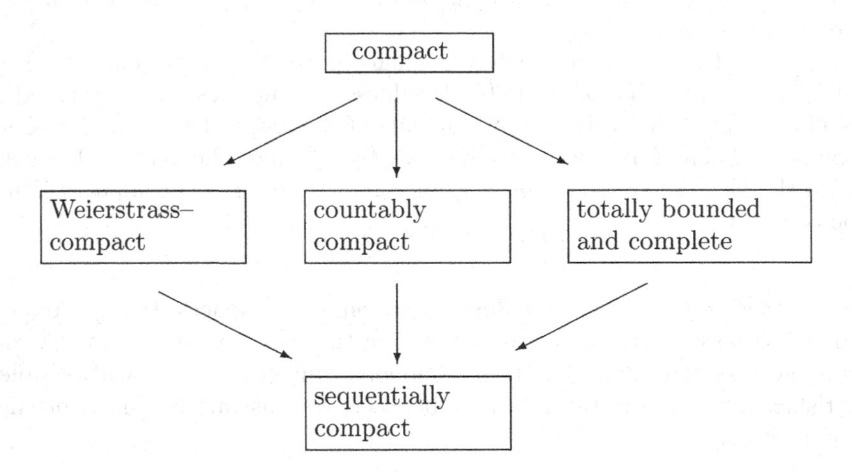

So it remains to be shown that **(4)** implies **(1)**. This will be done in 2 steps:

(4) $\Rightarrow$ **(5)** If X is sequentially compact, then X is complete. Assume that X fails to be totally bounded. Then there exists a positive real number r such that for each $n \in \mathbb{N}^+$ the set

$$X_n = \{(x_1, \ldots, x_n) \in X^n \mid i \neq j \Rightarrow d(x_i, x_j) \geq r\}$$

is non–empty. Under **CC** there exists an element (a_n) in $\prod_n X_n$. Consider $a_1, a_2, a_3, \ldots$ by concatenation as a sequence in X. This sequence contains arbitrary long strings $x_n, x_{n+1}, \ldots, x_{n+k}$ such that any two of its members have a distance $d(x_{n+i}, x_{n+j}) \geq r$. This fact immediately implies that (x_n) has a subsequence $(x_{\nu(n)})$ in which the distance of any two members is at least

[32] [BeHe98]

$\frac{1}{2}r$. Thus $(x_{\nu(n)})$ has no convergent subsequence, contradicting the sequential compactness of X.

(5) ⇒ (1) Let X be totally bounded and complete. Assume that $\mathcal{U}$ is an open cover of X without a finite subcover of X. For $x \in X$ and $r \in \mathbb{R}^+$, define $S(x,r) = \{y \in X \mid d(x,y) < r\}$. By total boundedness of X, for each $n \in \mathbb{N}^+$, the set A_n of all tuples $(x_1, \ldots, x_m)$ in X with $X = \bigcup_{i=1}^m S(x_i, \frac{1}{n})$ is non–empty. By **CC**, there exists an element (a_n) in $\prod\limits_n A_n$. Denote each a_n by $(x_1^n, x_2^n, \ldots x_{m(n)}^n)$, and define, via recursion, a sequence (y_n) in X as follows:

$$\begin{aligned}
&y_1 = x_k^1 \text{ where } k = \min\{i \mid \text{ no finite subset of } \mathcal{U} \text{ covers } S(x_i^1, 1)\}\\
&y_{n+1} = x_k^{n+1} \text{ where}\\
&k = \min\left\{i \mid \text{ no finite subset of } \mathcal{U} \text{ covers } S\left(x_i^{n+1}, \tfrac{1}{n+1}\right) \cap \bigcap_{\nu=1}^{n} S\left(y_\nu, \tfrac{1}{\nu}\right)\right\}.
\end{aligned}$$

Then (y_n) is a Cauchy sequence in X. By completeness, (y_n) converges to some point y in X. Thus there exists some $n \in \mathbb{N}^+$ and some $U \in \mathcal{U}$ with $S(y, \frac{1}{n}) \subseteq U$, providing a contradiction. Thus X is compact.

For the above result to hold, **CC** is not only sufficient but also necessary. More precisely:

Theorem 3.27. [33] *In the realm of pseudometric spaces the following conditions are equivalent:*

1. **CC**.
2. *Compact = sequentially compact.*
3. *Compact = totally bounded and complete.*
4. *Weierstrass–compact = totally bounded and complete.*
5. *Countably compact = totally bounded and complete.*
6. *Sequentially compact = totally bounded and complete.*
7. *Sequentially compact = countably compact.*

Proof. By Proposition 3.26, (1) implies all the above conditions. To show that each of these implies (1), assume that (1) fails. Then, by Theorem 2.12(3), also **PCC** fails, i.e., there exists a sequence (X_n) of non–empty sets such that each sequence in $\bigcup\limits_n X_n$ meets only finitely many X_n's.

Construct a pseudometric space with underlying set $X = \bigcup\limits_n (X_n \times \{n\})$ and distance function d, defined by

$$d((x,n),(y,m)) = \begin{cases} 0, & \text{if } n = m \\ 1, & \text{otherwise.} \end{cases}$$

Then (X,d) is sequentially compact, but neither totally bounded nor countably compact. Thus (2), (6), and (7) fail. Next, construct a pseudometric space with underlying set X as above and distance function a, defined by

33 [BeHe98]

$$a((x,n),(y,m)) = \left|\frac{1}{n} - \frac{1}{m}\right|.$$

Then (X, a) is complete and totally bounded, but fails to be countably compact. Thus (3) and (5) fail.

It remains to be shown that (4) implies (1). Assume that (1) fails, choose a sequence (X_n) as above, and construct a space (X, d) as above. Since (X, d) fails to be totally bounded, condition (4) implies that it fails to be Weierstrass–compact. So there exists an infinite subset A of X without an accumulation point in X. Define $A_n = A \cap (X_n \times \{n\})$ and $M = \{n \in \mathbb{N} \mid A_n \neq \emptyset\}$. Then $(A_m)_{m \in M}$ is a countable family of non–empty finite sets.

Assume that no sequence in $Z = \bigcup_{m \in M} A_m$ meets infinitely many A_m's. Then the pseudometric space (Z, b), defined by

$$b((x,n),(y,m)) = \left|\frac{1}{n} - \frac{1}{m}\right|$$

is complete and totally bounded, but not Weierstrass–compact, thus violating condition (4). Therefore there exists a sequence in Z which meets infinitely many A_m's and thus infinitely many X_m's, contradicting the choice of the X_n's. Thus (1) holds.

Observe that the first part of the proof of (4) $\Rightarrow$ (1) above also shows that the implication

Weierstrass–compact $\Rightarrow$ countably compact

implies **PCMC** and thus **CMC**. This implication is an equivalence, as the following result, presented here without proof, states:

Theorem 3.28. [34] *In the realm of pseudometric spaces the following conditions are equivalent:*

1. **CMC**.
2. *Weierstrass–compact $\Rightarrow$ countably compact.*
3. *Weierstrass–compact = compact.*

Proposition 3.29. [35] *In the realm of pseudometric spaces the following conditions are equivalent:*

1. **Fin**.
2. *Weierstrass–compact = sequentially compact.*

---

[34] [Ker2000]
[35] [BeHe98]

Proof. **(1) ⇒ (2)** Every Weierstrass–compact space is sequentially compact. For the converse, consider an infinite subset A of a sequentially compact space X. By (1), there exists an injective sequence (a_n) in A. By sequential compactness of X, some subsequence of (a_n) converges to some $x \in X$. Then x is an accumulation point of A. Thus X is Weierstrass–compact.

(2) ⇒ (1) If (1) fails, there exists an infinite, D–finite set X. Then the pseudometric space (X, d) defined by $d(x,y) = \begin{cases} 0, \text{ if } x = y \\ 1, \text{ if } x \neq y \end{cases}$ is sequentially compact, but not Weierstrass–compact. Thus (2) fails.

Finally let us sketch the situation for subspaces of $\mathbb{R}$. More details will be presented in Section 4.6.

Proposition 3.30. [36] *For subspaces X of $\mathbb{R}$, the following conditions are equivalent:*

1. *X is compact.*
2. *X is countably compact.*
3. *X is closed and bounded in $\mathbb{R}$.*
4. *X is sequentially compact and Lindelöf.*

Proof. The proof that (1), (2), and (3) are equivalent can be carried out as in **ZFC**. To show that (4) implies (1) consider two cases (cf. Theorem 3.16):
Case 1: CC($\mathbb{R}$) holds.
In this case, every sequentially compact space is easily seen to be closed and bounded in $\mathbb{R}$, thus compact.

Case 2: CC($\mathbb{R}$) fails.
In this case, every Lindelöf subspace of $\mathbb{R}$ is already compact (see Section 7.1).

Proposition 3.31. [37] *In the realm of subspaces of $\mathbb{R}$, the following conditions are equivalent:*

1. *Sequentially compact = compact.*
2. *Sequentially compact ⇒ closed.*
3. *Sequentially compact ⇒ bounded.*
4. *Complete ⇒ closed.*
5. *$\mathbb{R}$ is a sequential space.*

Proof. See Theorem 4.55 and Exercises to Section 4.6, E 4.

Theorem 3.32. *1. If there is no free ultrafilter on $\mathbb{R}$, then every subspace of $\mathbb{R}$ is ultrafilter–compact.*

2. If there is no free ultrafilter on $\mathbb{N}$, then $\mathbb{N}$ is Tychonoff–compact.

3. If $\mathbb{R}$ has an infinite, D–finite subset X, then the following hold:

[36] [Gut2003]

[37] [Gut2003]

a) $\mathbb{R}$ *has compact subspaces that fail to be Alexandroff–Urysohn–compact, e.g., the closed unit interval* $[0,1]$.
b) $\mathbb{R}$ *has subspaces that are Weierstrass–compact and complete, but fail to be Alexandroff–Urysohn–compact or compact, e.g.,* X *itself.*

Proof. **(1)** Immediate.

(2) Since $\mathbb{N}$ is discrete every subset of $\mathbb{N}$ is a zeroset. Thus in $\mathbb{N}$ every zero–ultrafilter is fixed and thus converges. By Exercise E 4 this implies that $\mathbb{N}$ is Tychonoff–compact.

(3) Let $\mathbb{R}$ have an infinite, D–finite subset X. Then every proper subset of X has a smaller cardinality than X. Consequently X has no complete accumulation point in $\mathbb{R}$. Hence neither X nor any subspace Y of $\mathbb{R}$ with $X \subseteq Y$ can be Alexandroff–Urysohn–compact. Let $f\colon \mathbb{R} \to (0,1)$ be some homeomorphism. Then $f[X]$ is an infinite, D–finite subset of $[0,1]$. Thus $[0,1]$ is not Alexandroff–Urysohn–compact.

Being D–finite, X contains no injective sequence. Thus X is complete. X is not closed, since otherwise one could construct an injective sequence in X. Thus X is not compact. However, X is Weierstrass–compact, for the assumption that there exists an infinite subset Y of X without accumulation point in X, leads to a contradiction as follows: for each $x \in X$ (except for the largest element of X, if X has one) there exists a largest half–open interval $I(x) = [x, r_x)$ in $\mathbb{R}$ with $[x, r_x) \cap Y = \emptyset$. Since these intervals $I(x)$ are pairwise disjoint and $\mathbb{Q}$ is dense in $\mathbb{R}$ and countable, it follows that the set of these intervals and thus X itself must be countable; a contradiction.

Exercises to Section 3.3:

E 1. [38] For topological spaces X, show the equivalence of the following conditions:
(1) X is compact.
(2) For each topological space Y the projection $\pi_Y\colon X \times Y \longrightarrow Y$ is a closed map.

E 2. [39] Show the equivalence of:
(1) X is a compact Hausdorff space.
(2) X is an H–closed regular space.

E 3. [40] For completely regular spaces X, show the equivalence of the following conditions:
(1) X is compact.
(2) In X, every zero–filter is fixed.

---

[38] [Her96]
[39] [Her96]
[40] [Her96]

(3) In the ring $C^*(X)$, every ideal is fixed.
(4) In the ring $C(X)$, every ideal is fixed.

E 4. [41] For completely regular spaces X, show the equivalence of the following conditions:
(1) X is Tychonoff–compact.
(2) The canonical map $X \to [0,1]^{C(X,[0,1])}$ is closed.
(3) In X, every zero–ultrafilter converges.
(4) In the ring $C^*(X)$, every maximal ideal is fixed.
(5) In the ring $C(X)$, every maximal ideal is fixed.

E 5. [42] Show that every Tychonoff–compact space is ultrafilter–compact.

E 6. [43] Show that, in case no free ultrafilters exist:
(1) Every space is ultrafilter–compact.
(2) Not every completely regular space is Tychonoff–compact.

E 7. [44] Show that in the realm of pseudometric spaces as well as in the realm of Hausdorff spaces the following conditions are equivalent:
(1) Compact = Alexandroff–Urysohn–compact.
(2) Ultrafilter–compact = Alexandroff–Urysohn–compact.
(3) **AC**.

E 8. Show the equivalence of:
(1) Every space with a finite topology is Alexandroff–Urysohn–compact.
(2) **AC**.

E 9. [45] Call a topological space X
- Lindelöf, if every open cover of X contains an at most countable subcover of X.
- w–Lindelöf (= weakly Lindelöf), if every open cover of X has an at most countable open refinement.
- vw–Lindelöf (= very weakly Lindelöf), if every open cover of X has an at most countable refinement.
- s–Lindelöf (= strongly Lindelöf), if for every extension Y of X, each open cover of X in Y contains an at most countable subcover of X.

Prove that:
(1) s–Lindelöf $\Rightarrow$ Lindelöf $\Rightarrow$ w–Lindelöf $\Rightarrow$ vw–Lindelöf.
(2) None of the above implications is an equivalence.
(3) In **ZFC** each of the above implications is an equivalence.
(4) Lindelöf = w–Lindelöf $\Leftrightarrow$ **CC**.
(5) Lindelöf = vw–Lindelöf $\Leftrightarrow$ **CC**.

[41] [Com68], [Sal74], [BeHe98].
[42] [Sal74]
[43] [Her96a]
[44] [How90]
[45] [Her2002]

(6) w–Lindelöf = vw–Lindelöf $\Leftrightarrow$ **CMC**.

(7) Equivalent are:

(a) Lindelöf = s–Lindelöf for T_1–spaces.

(b) $\mathbf{CC}(\mathbb{R})$ implies **CC**.

4

Disasters without Choice

Absence of choice
— in mathematics as in life —
may affect outcome.
S. Shelah and A. Soifer[1]

4.1 Finiteness

Without **AC**, *however, things become "sticky".*
J.L. Hickman[2]

Elementare Begriffe, wie Endlichkeit und Wohlordnung hängen jeweils vom gewählten System (Σ_1 oder Σ_2) ab; und es ist nicht ausgeschlossen, daß dieses Abhängen von wesentlichem Charakter ist: daß eine Menge a im System Σ_1 wohlgeordnet (bzw. endlich) zu sein scheint und sich im "feineren" System Σ_2 als nicht wohlgeordnet (bzw. unendlich) herausstellt.[3]
J. von Neumann

The concept of *finiteness*, defined as commonly done via natural numbers, is categorical, thus not problematic. If, however, the concept of being finite

[1] [ShSo2003]

[2] [Hic76]

[3] "Elementary concepts such as finiteness and well-order depend on the chosen system (Σ_1 or Σ_2) and it is conceivable that this dependence is essential: a set A may appear to be well-ordered (resp. finite) in system Σ_1, but turn out not to be well-ordered (resp. infinite) in the "finer" system Σ_2." [vNeu25]. Note that von Neumann's skepticism, expressed above, uses — independently of Tarski — the finiteness-definition that we present in 4.3.

is considered to be more fundamental than that of number, as strongly advocated, e.g., by Frege and Dedekind (and natural numbers are defined as the cardinals of finite sets), problems arise. First of all the concept of being finite loses its absoluteness. Secondly, how should finiteness be defined? There are several different descriptions, all equivalent to each other in **ZFC**, but not so in **ZF**. Historically the oldest definition is due to Dedekind (1888)[4].

It leads to disaster!

Definition 4.1. [5] *A set X is called* Dedekind–infinite *or just D*–infinite *provided that there exists a proper subset Y of X with $|X| = |Y|$; otherwise X is called* Dedekind–finite *or just D*–finite.

Proposition 4.2. *Equivalent are:*

1. *X is D–infinite.*
2. $|X| = |X| + 1$.
3. [6] $\aleph_0 \leq |X|$.

Proof. **(3) ⇒ (2)** Let $f\colon \mathbb{N} \to X$ be an injection, and let ∞ be an element, not contained in X. Then the map $g\colon X \to X \cup \{\infty\}$, defined by

$$g(x) = \begin{cases} \infty, & \text{if } x = f(0) \\ f(n), & \text{if } x = f(n+1) \\ x, & \text{otherwise} \end{cases},$$

is a bijection.

(2) ⇒ (1) Let ∞ be an element, not contained in X, and let $f\colon X \to X \cup \{\infty\}$ be a bijection. Then f^{-1}, restricted to X, is an injection from X onto the proper subset $X\backslash\{f^{-1}(\infty)\}$ of X.

(1) ⇒ (3) Let $f\colon X \to X$ be an injection onto a proper subset of X. Choose an element y in $X\backslash f[X]$ and define recursively a map $g\colon \mathbb{N} \to X$ by $g(0) = y$ and $g(n+1) = f(g(n))$. Then g is an injection.

Disaster 4.3. The following can happen:

1. [7] D–finite unions of D–finite sets may be D–infinite.
2. The power set of a D–finite set may be D–infinite.
3. A D–infinite set may be the image of a D–finite set.[8]

---

[4] [Ded1888]

[5] Cf. Definition 2.13.

[6] I.e., there exists an injection $\mathbb{N} \to X$. Here the natural numbers are supposed to have their familiar properties, being either defined axiomatically or as the cardinal numbers of finite sets (as defined in 4.4). The cardinal numbers of D–finite sets may contain "infinitely large" members and fail badly to satisfy the principle of induction, a real disaster!

[7] Contrast this with Exercise E 14a.

[8] Even worse: any $\aleph_\alpha$ (no matter how large) maybe the image of some D–finite set. See [Mon75].

Proof. Consider a model of **ZF** with the following property:

(F) There exists a sequence (X_n) of pairwise disjoint 2–element sets $X_n = \{x_n, y_n\}$ such that $X = \bigcup X_n$ is *D*–finite

(Such models exist[9]— another disaster). Then

(1) For each $x \in X$ consider the set $Y_x = \{x, n\}$, where n is the unique natural number with $x \in X_n$. Then $Y = \bigcup_{x \in X} Y_x$ is a *D*–finite union of *D*–finite sets, but $Y = \mathbb{N} \cup \bigcup_n X_n$ is *D*–infinite, since the map $f\colon \mathbb{N} \to Y$, defined by $f(n) = n$ is obviously injective.

(2) Though X is *D*–finite, the power set $\mathcal{P}X$ is *D*–infinite, since the map $f\colon \mathbb{N} \to \mathcal{P}X$, defined by $f(n) = \bigcup_{m \leq n} X_m$ is injective.

(3) Though X is *D*–finite, the map $f\colon X \to \mathbb{N}$, defined by

$$f(x) \text{ is the unique } n \in \mathbb{N} \text{ with } x \in X_n$$

is surjective.

The foregoing disasters show that — in the absence of **AC** — the above definition of *D–finiteness* is badly flawed. As satisfactory concept can however be obtained — as shown by Tarski (1924) — in the following way:

Definition 4.4. [10] *A set X is called* finite, *provided that each non–empty subset of $\mathcal{P}X$ contains a minimal element with respect to the inclusion order. Sets that are not finite are called* infinite.

Here follow some sample results:

Proposition 4.5. [11] *Equivalent are:*

1. *X is finite.*
2. *If $\mathfrak{A} \subseteq \mathcal{P}X$ satisfies*
   a) $\emptyset \in \mathfrak{A}$, and
   b) $A \in \mathfrak{A}$ and $x \in X$ imply $(A \cup \{x\}) \in \mathfrak{A}$,
   then $X \in \mathfrak{A}$.

Proof. **(1) ⇒ (2)** The collection $\mathfrak{B} = \{X \backslash A \mid A \in \mathfrak{A}\}$ has a minimal element B. Consequently $\mathfrak{A}$ has a maximal element $A = X \backslash B$. Thus (b) implies that $A = X$.

(2) ⇒ (1) Let $\mathfrak{A}$ be the set of all finite subsets of X. Since $\mathfrak{A}$ satisfies (a) and (b), (2) implies that $X \in \mathfrak{A}$. Thus X is finite.

---

9 See, e.g., Model A5 (N2(2) in [HoRu98]).

10 [Tar24a]. Cf. also [vNeu25].

11 [Tar24a]. Observe that by Proposition 4.5, the definition of finiteness as given in 4.4 is equivalent to the traditional definition that X is finite iff $|X| = n$ for some $n \in \mathbb{N}$.

Proposition 4.6. [12] *If X and Y are finite, so is $X \cup Y$.*

Proof. Let $\mathfrak{A}$ be a non–empty subset of $\mathcal{P}(X \cup Y)$. Then $\mathfrak{B} = \{A \cap X \mid A \in \mathfrak{A}\}$ contains a minimal element B, since X is finite; and $\mathcal{C} = \{A \cap Y \mid A \in \mathfrak{A}$ and $A \cap X = B\}$ contains a minimal element C, since Y is finite. Thus $B \cup C$ is a minimal element of $\mathfrak{A}$.

Proposition 4.7. [13] *Finite unions of finite sets are finite.*

Proof. Let $\mathfrak{M}$ be a finite set of finite sets. Consider

$$\mathfrak{A} = \{\mathfrak{B} \subseteq \mathfrak{M} \mid \bigcup \mathfrak{B} \text{ is finite}\}.$$

Then, in view of 4.6, $\mathfrak{A}$ satisfies the conditions (a) and (b) of 4.5(2). Thus $\mathfrak{M} \in \mathfrak{A}$, i.e., $\bigcup \mathfrak{M}$ is finite.

Proposition 4.8. [14] *If X is finite, then so is $\mathcal{P}X$.*

Proof. Consider $\mathfrak{A} = \{A \subseteq X \mid \mathcal{P}A \text{ is finite }\}$. Then, in view of 4.6, $\mathfrak{A}$ satisfies the conditions (a) and (b) of 4.5(2). Thus $X \in \mathfrak{A}$, i.e., $\mathcal{P}X$ is finite.

Proposition 4.9. [15] *Images of finite sets are finite.*

Proof. Let X be finite, and let $f\colon X \to Y$ be a surjection. If $\mathfrak{A}$ is a non–empty subset of $\mathcal{P}Y$, then $\mathfrak{B} = \{f^{-1}[A] \mid A \in \mathfrak{A}\}$ is a non–empty subset of $\mathcal{P}X$, and thus contains a minimal element B. Then $f[B]$ is a minimal element of $\mathfrak{A}$.

Let us return to Dedekind's definition of D–finiteness. How are the concepts of *finiteness* and *D–finiteness* related to each other?

Proposition 4.10. *Every finite set is D–finite.*

Proof. Assume that X is D–infinite. Then there exists an injection $f\colon \mathbb{N} \to X$. Consequently the collection $\mathfrak{A} = \{\{f(m) \mid m \geq n\} \mid n \in \mathbb{N}\}$ of subsets of X is non–empty, but contains no minimal element. Thus X is infinite.

The converse, however, is not true. I.e., there exist models of **ZF** in which there exist infinite, D–finite sets[16]

When do the two finiteness–concepts coincide? Precisely, if the disasters of 4.3 do not occur. The following Lemma will prepare the ground for the corresponding Theorem:

12 [Tar24a]
13 [Tar24a]
14 [Tar24a]
15 [Tar24a]
16 E.g., in Cohen's First Model A4 (M1 in [HoRu98]).

Lemma 4.11. *Equivalent are:*

1. $\aleph_0 \leq^* |X|$, *i.e., there exists a surjection* $X \to \mathbb{N}$.
2. $\mathcal{P}X$ *is D–infinite, i.e., there exists an injection* $\mathbb{N} \to \mathcal{P}X$.

Proof. **(1) ⇒ (2)** Let $f\colon X \to \mathbb{N}$ be a surjection. Then the map $g\colon \mathbb{N} \to \mathcal{P}X$, defined by $g(n) = f^{-1}(n)$, is an injection.

(2) ⇒ (1) Let $f\colon \mathbb{N} \to \mathcal{P}X$ be an injection. Define recursively a map $g\colon \mathbb{N} \to \mathcal{P}X$ such that the $g(n)$'s are non–void and pairwise disjoint: For $n \in \mathbb{N}$, assume that the $g(m)$'s are defined for all $m < n$ such that the set $\{f(k) \setminus \bigcup_{m<n} g(m) \mid k \geq n\}$ is infinite. Define

$$n^* = \min\{k \mid k \geq n \text{ and } f(k) \setminus \bigcup_{m<n} g(m) \neq \emptyset \neq (X \setminus f(k)) \setminus \bigcup_{m<n} g(m)\}.$$

In case $\{f(k) \setminus (f(n^*) \cup \bigcup_{m<n} g(m)) \mid k > n^*\}$ is infinite, define $g(n) = f(n^*) \setminus \bigcup_{m<n} g(m)$; otherwise define $g(n) = X \setminus (f(n^*) \setminus \bigcup_{m<n} g(m))$. The so defined $g\colon \mathbb{N} \to \mathcal{P}X$ has the required properties. Thus the map $h\colon X \to \mathbb{N}$, defined by $h(x) = \begin{cases} n, & \text{if } x \in g(n) \\ 0, & \text{if } x \notin \bigcup_{n\in\mathbb{N}} g(n) \end{cases}$, is a surjection.

Theorem 4.12. *Equivalent are:*

1. *Finite = D–finite.*
2. *D–finite unions of D–finite sets are D–finite.*
3. *Images of D–finite sets are D–finite.*
4. *The power set of each D–finite set is D–finite.*
5. *For each set* X *we have* $\aleph_0 \leq |X|$ *or* $|X| \leq \aleph_0$.

Proof. **(1) ⇒ (2)** Proposition 4.7.

(2) ⇒ (3) Let $f\colon X \to Y$ be a surjection with *D*–finite domain X. Then $Y = \bigcup_{x\in X} \{f(x)\}$ is a *D*–finite union of *D*–finite sets, thus *D*–finite.

(3) ⇒ (4) Assume that $\mathcal{P}X$ is *D*–infinite. Then, by Lemma 4.11, there exists a surjection $f\colon X \to \mathbb{N}$. Since $\mathbb{N}$ is *D*–infinite, (3) implies that X is *D*–infinite.

(4) ⇒ (1) It suffices to show that each infinite set X is *D*–infinite. The map $f\colon \mathbb{N} \to \mathcal{P}\mathcal{P}X$, defined by $f(n) = \{A \subseteq X \mid |A| = n\}$ is injective. Thus $\mathcal{P}\mathcal{P}X$ is *D*–infinite. Hence (4) implies that $\mathcal{P}X$ is *D*–infinite. Hence, by (4) again, X is *D*–infinite.

(1) ⇔ (5) Straightforward.

Observe that *D*–finite sets can possibly be quite "large". If X is *D*–finite and $\mathcal{P}X$ is *D*–infinite, then $\aleph_0 \leq^* |X|$ by 4.11. Moreover, Monro[17] has shown

[17] [Mon75]

that it is consistent to assume that for any $\aleph_\alpha$ (no matter how large) there exist D–finite sets X with $\aleph_\alpha \leq^* |X|$.

AC implies that finite = D–finite. **CC** suffices:

Proposition 4.13. *Under* **CC**, *finite* = *D–finite.*

Proof. It suffices to show that each infinite set X is D–infinite. If X is infinite then, for each $n \in \mathbb{N}$, the set X_n of all injective n–tuples $(x_1, \ldots, x_n)$ in X is non–empty. Thus, by **CC**, there exists an element $(y_n) \in \prod_n X_n$. Concatenation of the y_n's yields a sequence (x_n) in X with infinite range (precisely: if $y_n = (x_n^1, \ldots, x_n^n)$, then $x_{n\frac{(n+1)}{2}+k} = x_{n+1}^k$ for $n \in \mathbb{N}$ and $k \in \{1, \ldots, n+1\}$).

Cancellation of repeatedly occurring terms yields an injection $f \colon \mathbb{N} \to X$ (precisely: $f(n) = x_{\min\{k | x_k \notin \{f(m) | m<n\}\}}$). Thus X is D–infinite.

However, there are models[18] of **ZF** that satisfy the equation finite = D–finite, but fail to satisfy **CC**. For other models[19] the equation finite = D–finite even fails for subsets of $\mathbb{R}$.

Though in general the classes of all finite sets and of all D–finite sets are different, the first of these classes is completely determined by the latter.

Proposition 4.14. [20] *Equivalent are:*

1. *X is finite,*
2. *$\mathcal{PP}X$ is D–finite.*

Proof. **(1) ⇒ (2)** Immediate from Propositions 4.10 and 4.8.

(2) ⇒ (1) Assume X to be infinite. Then the map $f \colon \mathbb{N} \to \mathcal{PP}X$, defined by

$$f(n) = \{A \subseteq X \mid |A| = n\}$$

is injective. Thus $\mathcal{PP}X$ is D–infinite.

Definition 4.15. *Cardinal numbers of infinite, D–finite sets are called* Dedekind cardinals.

By Proposition 4.13, **CC** implies that there are no Dedekind cardinals. However, the next result, which we present without proof, shows that if there is at least one Dedekind cardinal, then there is a multitude of them with rather bizarre properties:

---

[18] E.g., Sageev's model (M6 in [HoRu98]).
[19] E.g., in Cohen's First Model A4 (M1 in [HoRu98]).
[20] [Tar24a]

Disaster 4.16. [21]

1. If there exists a Dedekind cardinal, then there is a set A of Dedekind cardinals which, supplied with its natural order, is order–isomorphic to $\mathbb{R}$. However, with respect to the order relation $\leq^*$ any two elements of A are comparable.
2. If there exist two non–comparable Dedekind cardinals, then there exists a countable set B of Dedekind cardinals that in its natural order forms an antichain.
3. In some models of **ZF** there are $2^{\aleph_0}$ non–comparable Dedekind cardinals.

Exercises to Section 4.1:

E 1. Show that a set X is D–infinite iff $\aleph_0 + |X| = |X|$.

E 2. Show that the following conditions are equivalent:
  a) X is finite.
  b) There exists an order relation $\leq$ on X such that $(X, \leq)$ and $(X, \geq)$ are well–ordered.
  c) There exists an order relation on X and each order–relation on X is a well–ordering.
  d) There exists an order relation on X and any two order relations on X are similar to each other (= order–isomorphic).
  e) In the lattice $\mathcal{P}X$ each ideal is principal.

E 3. [22] Investigate the relations between the following statements:
  a) X is finite.
  b) $\mathcal{P}X$ is D–finite.
  c) $(Y \subseteq X$ and $|X| \leq^* |Y|) \Rightarrow Y = X$.
  d) X is D–finite.
  e) $X = \emptyset$ or $|X| < 2 \cdot |X|$.
  f) $|X| \leq 1$ or $|X| < |X|^2$.
  g) $|X| \leq 1$ or there exists a subset Y of X with $|Y| < |X|$ and $|X \backslash Y| < |X|$.
  h) There exists a map $f \colon X \to X$ such that $\emptyset$ and X are the only subsets S of X with $f[S] \subseteq S$.

E 4. Show that finite products of finite sets are finite.

E 5. Show that the following conditions are equivalent.
  a) Finite = D–finite.
  b) If $\aleph_0 \leq^* |X|$, then $\aleph_0 \leq |X|$, (i.e., if there exists a surjection $f \colon X \to \mathbb{N}$, then there exists an injection $g \colon \mathbb{N} \to X$).
  c) If X and Y are disjoint infinite sets, then $X \cup Y$ is D–infinite.

---

[21] [Tar65], [Tru74].
[22] [Tar38], [Lev58], [HoYo89], [DCr2002].

E 6. [23] Show the equivalence of:

a) Finite = D–finite for subsets of $\mathbb{R}$.

b) $\mathbb{R}$ has no dense, D–finite subset.

[Cf. Exercises to Section 4.6, E 5.]

E 7. (a) Show that every non–empty set X with $2 \cdot |X| = |X|$ is D–infinite.

(b) Show that, if $a = 2 \cdot a$ for all infinite cardinals, then finite = D–finite.

E 8. Show the equivalence of:

a) **AC**.

b) X is finite iff for any well–ordering $\leq$ of X, the inverse order $\geq$ well–orders X, too.

E 9. Let $\mathcal{P}_{\text{fin}}X$ be the set of all finite subsets of X. Show the equivalence of:

a) $|X| = |\mathcal{P}_{\text{fin}}X|$ for each infinite set X.

b) **AC**.

E 10. [24] Show for infinite cardinals a:

a) If $a \leq b$ and b is a Dedekind cardinal, then so is a.

b) If a and b are Dedekind cardinals, then so are $a + b$ and $a \cdot b$.

c) a is a Dedekind cardinal iff a and $\aleph_0$ are incomparable.

d) a is a Dedekind cardinal iff $a + b = a + c$ implies $b = c$.

E 11. An *amorphous* set is an infinite set which has no infinite subset with an infinite complement.[25] Show that:

a) Each amorphous set is D–finite.

b) Under **OP** there are no amorphous sets.

c) If X is amorphous, then $X \uplus X$ is a non–amorphous D–finite set.

E 12. Show the equivalence of:

a) **AC**.

b) If $|A| < |A \cup B|$ and $|B| < |A \cup B|$, then $A \cup B$ is finite.

E 13. [26] Show the equivalence of:

a) Finite = D–finite.

b) $\prod_{i \in I} X_i \neq \emptyset$ for each family $(X_i)_{i \in I}$ of non–empty sets, indexed by a D–finite set I.

c) $\prod_{n \in \mathbb{N}} X_n \neq \emptyset$ for each sequence (X_n) of non–empty, D–finite sets.

E 14. Show that

a) D–finite unions of pairwise disjoint D–finite sets are D–finite.

b) Finite unions of D–finite sets are D–finite.

---

[23] [Bru82a]

[24] [Tar65]

[25] Amorphous sets exist in some **ZF**–models, e.g., in the basic Fraenkel's First Model A7 (N1 in [HoRu98]). Cf. also [Hic76].

[26] [DCr2002]

c) D–finite unions of finite sets may be D–infinite.

E 15. Show the equivalence of:
a) Finite = D–finite.
b) For each family $(X_i)_{i\in I}$, indexed by a D–finite set I, there exists a family $(Y_i)_{i\in I}$ of pairwise disjoint subsets Y_i of X_i with $\bigcup_{i\in I} Y_i = \bigcup_{i\in I} X_i$.

[Hint: Use Exercise E 14. above and Theorem 4.12.]

4.2 Disasters in Cardinal Arithmetic

What remains is to show that given two sets A and B, one is less than or equal to the other. If one thinks of this problem for two "arbitrary" sets, one sees the hopelessness of trying to actually define a map from one into the other. I believe that almost anyone would have a feeling of unease about this problem; namely that, since nothing is given about the sets, it is impossible to begin to define a specific mapping. This intuition is, of course, what lies behind the fact that it is unprovable in the usual Zermelo–Fraenkel set theory.

P. Cohen[27]

Many interesting and deep investigations on cardinal numbers become trivial if the axiom of choice is accepted.

J. Mycielski[28]

For cardinals a and b, the order relation $a \leq b$, their sum $a + b$, product $a \cdot b$, and power b^a are defined in **ZF** as in **ZFC**; so are finite sums $\sum_{\nu=0}^{n} a_\nu$ and finite products $\prod_{\nu=0}^{n} a_\nu$. But neither countable sums $\sum_{n\in\mathbb{N}} a_n$ nor countable products $\prod_{n\in\mathbb{N}} a_n$ can be defined as usual, since the following disaster can occur:

Disaster 4.17. [29] In some models of **ZF** there exist sequences $(A_n)_{n\in\mathbb{N}}$ and $(B_n)_{n\in\mathbb{N}}$ of sets such that $|A_n| = |B_n|$ for each $n \in \mathbb{N}$, but $|\biguplus_{n\in\mathbb{N}} A_n| \neq |\biguplus_{n\in\mathbb{N}} B_n|$ and $|\prod_{n\in\mathbb{N}} A_n| \neq |\prod_{n\in\mathbb{N}} B_n|$.

Proof. If **CC(2)** fails there exists a sequence $(A_n)_{n\in\mathbb{N}}$ of 2–element sets A_n with $\prod_{n\in\mathbb{N}} A_n = \emptyset$. Let $B_n = \{0,1\}$ for each $n \in \mathbb{N}$. Then:

1. $|A_n| = 2 = |B_n|$ for each $n \in \mathbb{N}$.

27 [Coh2002]
28 [Myc64]
29 $|A|$ denotes the cardinal number of the set A.

2. $|\prod_{n\in\mathbb{N}} A_n| = 0 \neq 2^{\aleph_0} = |\prod_{n\in\mathbb{N}} B_n|$.
3. $|\biguplus_{n\in\mathbb{N}} A_n| \neq \aleph_0 = |\biguplus_{n\in\mathbb{N}} B_n|$.

The unwelcome phenomenon, described above, has been illustrated by Russell by means of a sequence of pairs of boots, where the right and left boots of each pair are distinguishable, and a sequence of pairs of socks, where the right and left socks of each pair are indistinguishable; and thus in the latter case *"we cannot choose one out of each of an infinite number of pairs unless we have a* **rule** *of choice, and in the present case no rule can be found."*[30]

Even more dramatic is the fact that already in the finite case undesired phenomena may occur. E.g., cardinals may no longer be comparable.

Definition 4.18. [31] *Cardinals a and b are called* comparable *w.r.t. $\leq$ (resp. w.r.t. $\leq^*$) iff $a \leq b$ or $b \leq a$ (resp. $a \leq^* b$ or $b \leq^* a$).*

Disaster 4.19. It can happen that:

1. There are cardinals a and b with $a \leq^* b$ and $a \not\leq b$.
2. There exist cardinals a and b that are incomparable w.r.t. $\leq$.
3. There exist cardinals a and b that are incomparable w.r.t. $\leq^*$.
4. There exist cardinals a and b with $a \leq^* b$, $b \leq^* a$, and $a \neq b$.

Proof. **(1)** If in a model of **ZF** there exists an infinite, D–finite set X (see Section 4.1), then there exists (see Exercises to Section 4.1, E 5) a cardinal a with $\aleph_0 \leq^* a$ and $\aleph_0 \not\leq a$. [Hint: If $\mathcal{P}X$ is D–infinite, choose $a = |X|$, otherwise choose $a = |\mathcal{P}X|$.]

(2) Let X be a set that cannot be well–ordered and let $\aleph$ be the Hartogs–number of X. Then $\aleph \not\leq |X|$, by definition; and $|X| \not\leq \aleph$, since otherwise X would be well–orderable.

(3) See the following theorem.

(4) See Theorem 7.21(3).

Theorem 4.20. [32] *Equivalent are:*

1. *Any two cardinals are comparable w.r.t. $\leq$.*
2. *Any two cardinals are comparable w.r.t. $\leq^*$.*
3. **AC**.

Proof. **(1) ⇒ (2)** follows from the trivial fact that $a \leq b$ implies $a \leq^* b$.

(2) ⇒ (3) Let X be an arbitrary set. Then $|Y| \leq^* |X|$ for some set Y implies $|Y| \leq |\mathcal{P}X|$. Thus the Hartogs–number $\aleph$ of $\mathcal{P}X$ satisfies $\aleph \not\leq^* |X|$. By

---

[30] [Rus07] Actually, Russell writes of indistinguishable *boots*, not *socks*.

[31] By definition,
- $|A| \leq |B|$ iff there exists an injection $f : A \to B$,
- $|A| \leq^* |B|$ iff $A = \emptyset$ or there exists a surjection $g : B \to A$.

[32] [Har15] and [Sie58].

(2), this implies $|X| \leq^* \aleph$, i.e., (for $X \neq \emptyset$) there exists a surjection $f\colon \aleph \to X$. Define a map $g\colon X \to \aleph$ by $g(x) = \min f^{-1}(x)$. Then g is injective. Thus X is well–orderable. Thus (3) holds.

(3) $\Rightarrow$ (1) Let X and Y be arbitrary sets. By (3), X and Y can be well–ordered. As well–ordered sets they are either order–isomorphic or one of them is order–isomorphic to an initial segment of the other. Thus $|X| = |Y|$ or $|X| \leq |Y|$ or $|Y| \leq |X|$.

In view of the above theorem, sets of cardinals may fail to be linearly ordered, and thus fail to be well–ordered with respect to either the order relations $\leq$ or $\leq^*$. Does at least every cardinal have a direct successor?

W.r.t. $\leq$ the following two theorems, that we include without proof, present two sharply contrasting answers:

Theorem 4.21. [33] *Every cardinal a has a minimal successor w.r.t. $\leq$, namely*

- $a + 1$, *if a is finite,*
- $a + \aleph$, *if a is infinite and $\aleph$ is the Hartogs–number of a.*

As we will see in Theorem 7.22, a cardinal may have several minimal successors.

Theorem 4.22. [34] *Equivalent are:*

1. *Every cardinal has a smallest successor w.r.t. $\leq$.*
2. **AC**.

Moreover, in **ZF** cardinal arithmetic tumbles. Whereas in **ZFC**, addition and multiplication, restricted to infinite cardinals a and b, are trivial operations satisfying

$$a \cdot b = a + b = \max\{a, b\}$$

hence in particular

$$a^2 = 2 \cdot a = a,$$

in **ZF** addition and multiplication are as simple as above just for Alephs, but no longer so for arbitrary infinite cardinals.

Lemma 4.23. [35] *If $a \cdot \aleph = a + \aleph$ for some cardinal a and some Aleph $\aleph$, then a and $\aleph$ are comparable, w.r.t. $\leq$.*

Proof. Let A be a set with $|A| = a$, let W be a well–ordered set, disjoint from A, with $|W| = \aleph$, and let $f\colon A \times W \to A \cup W$ be a bijection. Consider $\bar{A} = f^{-1}[A]$ and $\bar{W} = f^{-1}[W]$. Then $|\bar{A}| = a$ and $|\bar{W}| = \aleph$.

---

33 [Tar54]
34 [Tar54]
35 [Tar24]

Case 1: There exists $\bar{a} \in A$ with $(\{\bar{a}\} \times W) \subseteq \bar{A}$.
Then $\aleph = |W| = |\{\bar{a}\} \times W| \leq |\bar{A}| = a$, thus $\aleph \leq a$.

Case 2: For each $a \in A$, $W_a = \{w \in W \mid (a, w) \in \bar{W}\} \neq \emptyset$.
Then each set W_a has a smallest element w_a. Hence:

$$a = |A| = |\{(a, w_a) \mid a \in A\}| \leq |\bar{W}| = \aleph,$$

thus $a \leq \aleph$.

Theorem 4.24. [36] *Equivalent are:*

1. $a^2 = a$ *for all infinite cardinals* a.
2. $a \cdot b = a + b$ *for all infinite cardinals* a *and* b.
3. **AC**.

Proof. **(1) ⇒ (2)** $a + b \leq a \cdot b \leq a^2 + 2ab + b^2 = (a + b)^2 = a + b$. Thus $a + b = a \cdot b$.

(2) ⇒ (3) Let A be an infinite set with $|A| = a$, and let $\aleph$ be the Hartogs–number of A. By (2), $a \cdot \aleph = a + \aleph$. Thus by Lemma 4.23, a and $\aleph$ are comparable w.r.t. $\leq$. Since $\aleph \not\leq a$, by definition, we conclude that $a < \aleph$. Thus A is well–orderable. This implies (3).

(3) ⇒ (1) Let A be an infinite set with $|A| = a$. By (3), A has a countable infinite subset N. Obviously, there exists a bijection $f \colon N \to N^2$. Consider the set $\mathfrak{M}$ of all pairs (M, g), where $N \subseteq M \subseteq A$ and $g \colon M \to M^2$ is a bijection. Order $\mathfrak{M}$ by

$$(M, g) \leq (K, h) \Leftrightarrow (M \subseteq K \text{ and } g \text{ is a restriction of } h).$$

Then, in the ordered set $(\mathfrak{M}, \leq)$, each chain has an upper bound. Thus (3), via Zorn's Lemma, implies that $(\mathfrak{M}, \leq)$ has a maximal element (B, g). Consider, $C = A \setminus B$, $|B| = b$, and $|C| = c$.

Case 1: $c \leq b$.
Then $a \leq a^2 = (b^2 + b \cdot c + c \cdot b + c^2) \leq 4b^2 = 4b \leq b^2 = b \leq a$. Thus $a = a^2$.

Case 2: $b \leq c$.
Then there exists a subset D of C with $|D| = b$. Consider $E = (B \cup D)^2 \setminus B^2$ and $e = |E|$. Then $b = |D| \leq |D^2| \leq e = 3b^2 = 3b \leq b^2 = b$.

Thus $e = b$. Consequently there exists a bijection $h = D \to E$. Thus the map $k \colon (B \cup D) \to (B \cup D)^2$, defined by $k(x) = \begin{cases} g(x), \text{ if } x \in B \\ h(x), \text{ if } x \in D \end{cases}$, is a bijection that extends $g \colon B \to B^2$. This contradicts the maximality of (B, g). Thus Case 2 cannot occur.

[36] [Tar24]. For a historical discussion of this result see [Dei2005].

The above theorem, due to Tarski, has a remarkable history concerning its publication. Moore[37] relates in his book, p. 215, the following story communicated to him by Tarski himself:

> *"Before these results appeared in* Fundamenta Mathematica *in 1924, Tarski sent Lebesgue a note showing that (4.3.2)* [i.e., condition (1) of the above theorem] *is equivalent to the Axiom, and asked him to submit it to the* Comptes Rendus *of the Paris Academy of Sciences. Lebesgue returned the note on the grounds that he opposed the Axiom, but suggested sending it to Hadamard. When Tarski did as Lebesgue advised, Hadamard also returned the note — saying that, since the Axiom was true, what was the point of proving it from (4.3.2)?"*

Exercises to Section 4.2:

E 1. Show that, for cardinals a, b and Alephs $\aleph$, the following hold:
a) $a \leq b \Rightarrow a \leq^* b$.
b) $a \leq \aleph \Leftrightarrow a \leq^* \aleph$.
c) $a \leq^* b \Rightarrow a \leq 2^b$.
d) $\aleph_0 \leq^* a \Leftrightarrow \aleph_0 \leq 2^a$ [cf. Lemma 4.11].
e) $a \leq^* b \Rightarrow 2^a \leq 2^b$.

E 2. Show that, for cardinals a, b, the inequalities $a \leq b \leq a$ imply $a = b$.
[Hint: Use the proof of Theorem 5.24.]

E 3. [38] Let a be the cardinal number of the set of all well–order relations on $\mathbb{N}$. Show that
a) $a \leq 2^{\aleph_0}$.
b) $2^{\aleph_0} \leq a$.
c) $a = 2^{\aleph_0}$.
[Hint: (b) Consider all well–order relations on $\mathbb{N}$ for which the subset of even numbers and the subset of odd numbers occur each in their natural order.
(c) Use E 2. above.]

E 4. [39] Show that $\aleph_1 \leq^* 2^{\aleph_0}$.
[Hint: Use E 3. above.]

E 5. Show that $|\mathbb{N}^{\mathbb{N}}| = |\mathbb{R}^{\mathbb{N}}| = |\mathbb{R}^n| = |\mathbb{R}| = |\mathcal{P}\mathbb{N}| = |\mathcal{P}\mathbb{Q}| = 2^{\aleph_0}$ for each $n \in \mathbb{N}^+$.

---

37 [Moo82]
38 [Chu27]
39 [Chu27]

E 6. [40] Show that for cardinals a, the following conditions are equivalent:
a) $a = 2 \cdot a$.
b) $a = \aleph_0 \cdot a$.

E 7. [41] Show that the following conditions are equivalent:
a) $a = 2 \cdot a$ for infinite cardinals.
b) $a + b$ is the least upper bound, w.r.t. $\leq$ of a and b for infinite cardinals a and b.

4.3 Disasters in Order Theory

Linearly ordered sets may have unfamiliar properties (cf. Exercise E 1).

Disaster 4.25. The following can happen in a linearly ordered set, even in a subset X of $\mathbb{R}$:

1. X contains no decreasing sequence, but fails to be well–ordered.
2. X is infinite but contains neither a decreasing[42] nor an increasing[42] sequence.
3. X is non–empty and without a largest element, but contains no increasing sequence.

Proof. Let X be an infinite, D–finite subset of $\mathbb{R}$, supplied with its natural order[43]. Then X contains neither an increasing nor a decreasing sequence. X is not well–ordered, since otherwise we could define an increasing sequence (x_n) via recursion by

$$x_n \text{ is the smallest element of } X \setminus \{x_m \mid m < n\}.$$

In case X has a largest element, there exists a finite set F such that $Y = X \setminus F$ is an infinite, D–finite set without a largest element, since otherwise we could define a decreasing sequence (x_n) in X via recursion by:

$$x_n \text{ is the largest element of } X \setminus \{x_m \mid m < n\}.$$

Disaster 4.26. A partially ordered set may have neither a maximal chain nor a maximal antichain.

---

40 [HaHo70]

41 [Haeu83]

42 A sequence (x_n) in an ordered set is called *decreasing* (resp. *increasing*) provided that $x_{n+1} < x_n$ (resp. $x_n < x_{n+1}$) for each $n \in \mathbb{N}$.

43 Such sets exist in certain models of **ZF**, e.g. in Cohen's First Model A4 (M1 in [HoRu98]).

Proof. Immediate from Theorems 2.2 and 2.4.

Next, let us turn our attention to the question whether lattices have maximal filters. For convenience we adopt the following slightly restricted definition of lattices[44].

Definition 4.27. *A* lattice *is a partially ordered set L in which each finite subset F has an infimum,* $\inf F$, *and a supremum,* $\sup F$, *(in particular L has a smallest element, $0 = \sup \emptyset$, and a largest element, $1 = \inf \emptyset$) and such that $0 \neq 1$.*

Definition 4.28. *A lattice L is called*

1. distributive, *provided it satisfies the equation $x \wedge (y \vee z) = (x \wedge y) \vee (x \wedge z)$ for all x, y, and z (and thus also the equation $x \vee (y \wedge z) = (x \vee y) \wedge (x \vee z)$).*
2. complete, *provided that each of its subsets has a supremum (and thus also an infimum).*
3. *a* powerset–lattice, *provided L is isomorphic to the lattice of all subsets of some non–empty set,*
4. *an* open lattice, *provided that L is isomorphic to the lattice $\tau(X)$ of all open sets of some non–empty topological space,*
5. *a* closed lattice, *provided that L is isomorphic to the lattice $\gamma(X)$ of all closed sets of some non–empty topological space X.*

Definition 4.29. *1. A subset F of a lattice L is called a* filter *in L iff the following two conditions are satisfied:*
(a) $1 \in F$ and $0 \notin F$.
(b) $(x \wedge y) \in F$ iff ($x \in F$ and $y \in F$).
2. *A filter F in L is called* maximal *iff L has no properly larger filter than F.*
3. *A filter F in L is called* prime *iff it satisfies the condition*
(c) $(x \vee y) \in F \Leftrightarrow (x \in F$ or $y \in F)$.
4. *A filter (resp. maximal filter) in the powerset–lattice of a set X is also called a* filter *(resp.* ultrafilter*) on X.*

Dual concepts: ideal, maximal ideal, prime ideal.

Observe that for distributive lattices every maximal filter is prime. However, the lattice

```
      o
    / | \
   o  o  o
    \ | /
      o
```

has precisely 4 filters, 3 of which are maximal, but none is prime.

In the lattice $\tau(\mathbb{R})$ of open sets in $\mathbb{R}$, for each $x \in \mathbb{R}$ the filter $F(x) = \{A \in \tau(\mathbb{R}) \mid x \in A\}$ is prime but not maximal. For Boolean algebras, in particular for powerset–lattices, a filter is prime iff it is maximal.

---

[44] By standard terminology our lattices would have to be called *non–trivial bounded* lattices.

Do lattices have maximal filters? In the case of powerset–lattices of the form $\mathcal{P}(X)$ the obvious answer is "yes", since for each $x \in X$ the set $\overset{\bullet}{x} = \{F \subseteq X \mid x \in F\}$ is an ultrafilter on X. Ultrafilters $\mathcal{F}$ of this simple type are called *fixed* (since $\bigcap \mathcal{F} \neq \emptyset$), all others are called *free* (since their intersection is empty). Are there any free ultrafilters on X? For finite X, obviously not. However, for infinite X there are $2^{2^{|X|}}$ free ultrafilters on X — provided that we work in **ZFC**. However, in **ZF** the situation is very different:

Disaster 4.30. The following might happen:

1. There are no free ultrafilters.
2. There are free ultrafilters on some sets, but there is none on $\mathbb{N}$.
3. There are free ultrafilters on every infinite set, but not every filter $\mathcal{F}$ on a set X can be enlarged to an ultrafilter on X.
4. There are sets with precisely one free ultrafilter.

Proof. **(1)**, **(2)** see [Bla77] or Pincus–Solovay's Model A6 (M27 in [HoRu98]).
(3) is true in Fraenkel's First Model A7 (N1 in [HoRu98]).
(4) See Exercise E 4.

What about the existence of maximal filters in one of the wider classes of lattices defined above? In the case of all lattices, the following proposition provides an answer.

Proposition 4.31. [45] *Equivalent are:*

1. *Every lattice has a maximal filter.*
2. **AC**.

Proof. **(1)** $\Rightarrow$ **(2)** Let $(X_i)_{i \in I}$ be a family of non–empty sets. Consider the set of all pairs (J, x) with $J \subseteq I$ and $x \in \prod_{j \in J} X_j$, ordered by

$$(J, x) \leq (K, y) \text{ iff } (J \subseteq K \text{ and } x \text{ is the restriction of } y \text{ to } J).$$

By adding a largest element 1, a lattice L results. By (1), the dual lattice has a maximal filter, thus L itself has a maximal ideal M. For (J, x) and (K, y) in M, the inequality $(J, x) \vee (K, y) \neq 1$ implies that $x_i = y_i$ for each $i \in (J \cap K)$. Thus the union of all the first components of members of M is a subset K of I, and the union of all the second components of members of M is an element x of $\prod_{k \in K} X_k$. Maximality of M implies $K = I$. Thus $x \in \prod_{i \in I} X_i$.
(2) $\Rightarrow$ **(1)** Immediate via Zorn's Lemma.

There is a deeper result: In condition (1) of the above proof the collection of all lattices can be reduced considerably.

45 [Sco54]

Theorem 4.32. [46] *Equivalent are:*

1. *Every lattice has a maximal filter.*
2. *Every complete lattice has a maximal filter.*
3. *Every distributive lattice has a maximal filter.*
4. *Every closed lattice has a maximal filter.*
5. **AC**.

Proof. In view of Proposition 4.31 and the fact that every closed lattice is complete and distributive it suffices to show that (4) implies (5):

(4) ⇒ (5) Let $(X_i)_{i\in I}$ be a family of non–empty sets. For each $i \in I$, add a new element ∞ to X_i, and consider $Y_i = X_i \cup \{\infty\}$ as a topological space whose collection of closed sets is

$$\gamma(X_i) = \{Y_i\} \cup \{A \subseteq X_i \mid A \text{ finite}\}.$$

Then the product space $Y = \prod_{i\in I} Y_i$ is non–empty, hence its lattice $\gamma(Y)$ of closed sets contains a maximal filter $\mathcal{F}$. For each $i \in I$ denote the i–th projection by $\pi_i : Y \to Y_i$, and define $\mathcal{F}_i = \{A \in \gamma(X_i) \mid \pi_i^{-1}[A] \in \mathcal{F}\}$. Then each $\mathcal{F}_i$ is a prime filter in $\gamma(X_i)$. Define $J = \{i \in I \mid \mathcal{F}_i = \{Y_i\}\}$ and $K = I \setminus J$. For each $k \in K$ there exists a unique element $x_k \in X_k$ with $\{x_k\} \in \mathcal{F}_k$. Thus $(x_k)_{k\in K} \in \prod_{k\in K} X_k$. It remains to be shown that $K = I$, i.e., $J = \emptyset$. Define $y = (y_i)_{i\in I} \in Y$ by

$$y_i = \begin{cases} x_i, & \text{if } i \in K \\ \infty, & \text{if } i \in J \end{cases}.$$

Consider $i \in I$ and a neighborhood U of y_i in Y_i. Then $\pi_i^{-1}[U]$ meets every member of $\mathcal{F}$. (This is obvious for $i \in K$. It holds for $i \in J$, since otherwise there would exist some finite subset A of X_i in $\mathcal{F}_i$, contradicting the definition of J.) Since $\mathcal{F}$ is prime, this implies that every neighborhood of y meets every member of $\mathcal{F}$. Since $\mathcal{F}$ consists of closed sets only, this implies that $y \in \bigcap \mathcal{F}$ and hence $\mathrm{cl}\{y\} \subseteq \bigcap \mathcal{F}$ is the product of the closures of its components[47], i.e., $\mathrm{cl}\{y\} = \prod_{i\in I} \mathrm{cl}_i\{y_i\} = \prod_{i\in I} Z_i$, with

$$Z_i = \begin{cases} \{x_i\}, & \text{if } i \in K \\ Y_i, & \text{if } i \in J \end{cases}.$$

In view of the maximality of $\mathcal{F}$ this implies $J = \emptyset$, since otherwise for a fixed $j \in J$ and a fixed $x \in X_j$ the filter in $\gamma(Y)$, generated by the set $\pi_j^{-1}(x)$, would be properly larger than $\mathcal{F}$. Thus $y \in \prod_{i\in I} X_i$.

[46] [Kli58], [Ban61], [KeTa99], [Her2003].
[47] where $\mathrm{cl}\{y\}$ is the closure of $\{y\}$ in Y

Corollary 4.33. *Equivalent are:*

1. *In every lattice each filter can be enlarged to a maximal one.*
2. *In every closed lattice every filter can be enlarged to a maximal one.*

Observe that the space Y, constructed in the proof of Theorem 4.32 is a T_0–space, but fails to be a T_1–space. In fact, for non–empty T_1–spaces X the corresponding closed lattices $\gamma(X)$ always have maximal closed filters, namely $\overset{\bullet}{x} = \{A \in \gamma(X) \mid x \in A\}$, for each $x \in X$.

The open lattices $\tau(X)$ are just the duals of the corresponding closed lattices $\gamma(X)$. So one might expect similar results. Surprisingly, however, the statement that each open lattice contains a maximal filter is weaker than **AC**.

Definition 4.34. *Let X be a topological space and $A \in \tau(X)$. Then*

$$A^* = \mathrm{int}(X \setminus A),$$

the interior of the complement of A, is called the pseudocomplement *of A.*

Lemma 4.35. *Let X be a topological space, $A \in \tau(X)$, and $\mathcal{F}$ a filter in $\tau(X)$. Then the following hold:*

1. *A^* is the largest element of $\tau(X)$ that misses A.*
2. *$\mathcal{F}$ is maximal iff the following condition is satisfied:*
   (*) For each $A \in \tau(X)$, either $A \in \mathcal{F}$ or $A^* \in \mathcal{F}$.

Proof. **(1)** is trivial.

(2) Let $\mathcal{F}$ be maximal and let A be an element of $\tau(X)$ that does not belong to $\mathcal{F}$. Then, by maximality of $\mathcal{F}$, there exists a member F of $\mathcal{F}$ with $F \cap A = \emptyset$. Thus $F \subseteq A^*$. This implies $A^* \in \mathcal{F}$.

Now, let $\mathcal{F}$ satisfy the condition (*) and let $\mathcal{G}$ be a filter in $\tau(X)$ with $\mathcal{F} \subseteq \mathcal{G}$. For $G \in \mathcal{G}$ we have either $G \in \mathcal{F}$ or $G^* \in \mathcal{F}$. The latter case $G^* \in \mathcal{F}$ cannot happen, since otherwise G and G^*, and thus $\emptyset = G \cap G^*$, would belong to $\mathcal{G}$. Thus $G \in \mathcal{F}$, hence $\mathcal{G} \subseteq \mathcal{F}$, hence $\mathcal{G} = \mathcal{F}$.

Theorem 4.36. [48] *Equivalent are:*

1. *Every open lattice has a maximal filter.*
2. *In every open lattice, every filter can be enlarged to a maximal one.*
3. **UFT**, *the Ultrafilter Theorem.*

Proof. **(1) ⇒ (2)** Let X be a topological space, and let $\mathcal{F}$ be a filter in $\tau(X)$. Let Y be the topological space, whose underlying set (also denoted by Y) consists of all filters in $\tau(X)$ that enlarge $\mathcal{F}$, and whose topology consists of all subsets of Y that contain with any element $\mathcal{G}$ all elements of Y that enlarge $\mathcal{G}$. By (1), there exists a maximal filter $\mathfrak{M}$ in $\tau(Y)$. For each $A \in \tau(X)$ define

48 [Rhi2002]

$$U(A) = \{\mathfrak{G} \in Y \mid A \in \mathcal{G}\}.$$

Then $\mathcal{U} = \{A \in \tau(X) \mid U(A) \in \mathcal{M}\}$ is an element of Y. Maximality of $\mathcal{U}$ will follow, by 4.35 from the fact that for each $A \in \tau(X)$ either A or its pseudocomplement A^* belongs to $\mathcal{U}$. To establish the latter fact, observe first that if $A \notin \mathcal{U}$, i.e., if $U(A) \notin \mathcal{M}$, maximality of $\mathcal{M}$ implies that there is some $M \in \mathcal{M}$ with $M \cap U(A) = \emptyset$. Thus $A \notin \mathcal{G}$ for each $\mathcal{G} \in M$. This implies $A^* \in \mathcal{G}$ for each $\mathcal{G} \in M$, since $A^* \notin \mathcal{G} \in M$ would imply that each member of $\mathcal{G}$ would meet A, hence there would exist an enlargement $\mathcal{H}$ of $\mathcal{G}$ — hence a member $\mathcal{H}$ of M — with $A \in \mathcal{H}$; a contradiction.
Consequently $M \subseteq U(A^*)$. This implies $U(A^*) \in \mathcal{M}$, thus $A^* \in \mathcal{U}$.

(2) ⇒ (3) Immediate, since (3) is the restriction of (2) to discrete spaces.

(3) ⇒ (1) Let X be a non–empty topological space (whose underlying set we also denote by X). Consider the filter $\mathcal{F} \subseteq \tau(X)$ consisting of all $F \in \tau(X)$ that are dense in the space X. Then F can be enlarged to a filter $\mathcal{G}$ on the set X, and, by (3), we may assume that $\mathcal{G}$ is an ultrafilter on X. Thus $\mathcal{H} = \mathcal{G} \cap \tau(X)$ is a prime filter in $\tau(X)$ that enlarges $\mathcal{F}$. For each $A \in \tau(X)$ the set $A \cup A^*$ is dense in X and thus belongs to $\mathcal{F}$, hence to $\mathcal{H}$. Since $\mathcal{H}$ is prime, this implies that $A \in \mathcal{H}$ or $A^* \in \mathcal{H}$. By 4.35, this implies that $\mathcal{H}$ is maximal in $\tau(X)$.

For the following result we regard **2** as a Boolean lattice with underlying set $\{0,1\}$ and $0 < 1$. Obviously, a subset F of a Boolean algebra B is a prime filter in B iff the map $f\colon B \to \mathbf{2}$, defined by $f(x) = \begin{cases} 1, \text{ if } x \in F \\ 0, \text{ otherwise} \end{cases}$, is a Boolean homomorphism.

Theorem 4.37. [49] *Equivalent are:*

1. **PIT** *the Prime Ideal Theorem: every Boolean lattice has a maximal filter*[50].
2. *In a Boolean lattice, every filter can be enlarged to a maximal one.*
3. **UFT**, *the Ultrafilter Theorem.*
4. *Products of compact Hausdorff spaces are compact.*
5. *Products of finite discrete spaces are compact.*

Proof. **(1) ⇒ (2)** Let $\mathcal{F}$ be a filter in a Boolean lattice B. Let $f\colon B \to B/\mathcal{F}$ the natural map from B onto the corresponding quotient space of B, with $f^{-1}(1) = \mathcal{F}$. By (1), there exists a maximal filter $\mathcal{G}$ in $B/\mathcal{F}$. Then $f^{-1}[\mathcal{G}]$ is a maximal filter in B that enlarges $\mathcal{F}$.

(2) ⇒ (3) Immediate, since (3) is the restriction of (2) to powerset–lattices.

---

[49] [RuSc54], [LoRy55], [Ban79].

[50] Since the concept of Boolean lattice is self–dual, and since in Boolean lattices the concepts of maximal and prime filters coincide, the above formulation is obviously equivalent to the one given in Section 2.2.

(3) ⇒ (4) By Theorem 3.22, (3) implies that a topological space X is compact iff every ultrafilter in X converges. Let $(X_i)_{i\in I}$ be a family of compact Hausdorff spaces, let $X = \prod_{i\in I} X_i$ be their product, and let $\mathcal{U}$ be an ultrafilter on X. Then, for each $i \in I$, the set $\mathcal{U}_i = \{A \subseteq X_i \mid \pi_i^{-1}[A] \in \mathcal{U}\}$ is an ultrafilter on X_i (where π_i denotes the i–th projection). Since X_i is compact and Hausdorff, $\mathcal{U}_i$ converges to a unique element x_i in X_i. Thus $\mathfrak{U}$ converges to $x = (x_i)_{i\in I}$.

(4) ⇒ (5) Trivial.

(5) ⇒ (1) Let B be a Boolean lattice. Consider the set $\mathcal{A}$ of all finite Boolean subalgebras of B. For $A \in \mathcal{A}$ the set X_A of all Boolean homomorphisms from A to $\mathbf{2}$ is non–empty (since obviously each finite Boolean lattice has a maximal filter). Consider each X_A as a discrete topological space. By (5), the product space $X = \prod_{A\in\mathcal{A}} X_A$ is compact and non–empty[51]. For any pair (A, B) of elements of $\mathfrak{A}$ with $A \subseteq B$ the set

$$C(A,B) = \{(x_C) \in X \mid x_A \text{ is the restriction of } x_B \text{ to } A\}$$

is closed in X, and the sets $C(A, B)$ have the finite intersection property. Since X is compact, there exists an element $x = (x_C)$ in the intersection of all $C(A, B)$'s. For $a \in B$, let $B(a)$ be the finite subalgebra of B generated by a. Then the map $f : B \to \mathbf{2}$, defined by $f(a) = x_{B(a)}$ is a Boolean homomorphism. Thus $f^{-1}(1)$ is a maximal filter in B.

Finally, let us investigate the question whether or not every set can be nicely ordered. We know already that it may not be possible to well–order some set X. But it is easily seen that each set can be partially ordered, even lattice–ordered, provided it has at least 2 elements (see Exercise E 7). Can it be linearly ordered?

Disaster 4.38. It can happen that certain sets cannot be linearly ordered.

The next result provides some indication of what may go wrong.

Proposition 4.39. [52] *Each of the following conditions implies all subsequent ones:*

1. **UFT**.
2. **OEP**, *the* **Order Extension Principle**: *for each partially ordered set (X, R) there exists a linear order relation S on X with $R \subseteq S$.*
3. **OP**, *the* **Ordering Principle**: *each set can be linearly ordered.*
4. **AC**(fin): *Products of non–empty, finite sets are non–empty.*

---

[51] The non–emptiness follows immediately from the fact that the product $\prod_{A\in\mathcal{A}} Y_A$ of the discrete spaces $Y_A = X_A \cup \{\infty\}$ is compact.

[52] D. Scott, as quoted from [Jec73].

Proof. **(1) ⇒ (2)** Let (X, R) be an partially ordered set. For each finite subset F of X let X_F be the set of all linear order relation S on F with $(R \cap F^2) \subseteq S$. Then each X_F, with $F \in \mathcal{P}_{\text{fin}}(X)$, is a non–empty finite set (cf. Exercise E 9.(2)). Consider it as a discrete topological space. Then, by (1), the space $Y = \prod_{F \in P_{\text{fin}}(X)} X_F$ is — according to Theorem 4.37 — a non–empty compact space. For any pair (E, F) of finite subsets of X define a set

$$A_{(E,F)} = \{(S_G)_{G \in P_{\text{fin}}(X)} \mid S_E \cap (E \cap F)^2 = S_F \cap (E \cap F)^2\}.$$

Then the sets $A_{(E,F)}$ are non–empty and closed in Y; and they have the finite–intersection property. Thus their intersection contains an element $(S_G)_{G \in P_{\text{fin}}(X)}$, which implies that $S = \bigcup_{G \in P_{\text{fin}}(X)} S_G$ is a linear order relation on X with $R \subseteq S$.

(2) ⇒ (3) By (2), the partial order relation on X defined by $x \leq y \Leftrightarrow x = y$ can be extended to a linear order relation on X.

(3) ⇒ (4) Let $(X_i)_{i \in I}$ be a family of non empty, finite sets. By (3), $X = \bigcup_{i \in I} X_i$ can be linearly ordered. Thus each X_i has a smallest element x_i in X. Consequently $(x_i) \in \prod_{i \in I} X_i$.

Proposition 4.40. [53] *The Kinna–Wagner Selection Principle,* **KW**, *implies the Ordering Principle,* **OP**.

Proof. Let X be a set. Consider the collection $\mathcal{P}_2(X)$ of all subsets A of X with $|A| \geq 2$. By **KW**, there exists a family $(A_Y)_{Y \in \mathcal{P}_2(X)}$ of non–empty, proper subsets A_Y of X. Denote $X \setminus A_Y$ by B_Y. Consider the set Z of all linear preorder relations R on X. For each R in Z and each x in X consider the component $[x]_R = \{y \in X \mid xRy \text{ and } yRx\}$ of x in (X, R). Let $\mathcal{K}_R$ be the set of all components $[x]_R$ of (X, R) with at least two elements. Let $\aleph$ be the Hartogs–number of Z, and define, via transfinite recursion, a map $f \colon \aleph \to Z$ by

1. $f(0) = X \times X$,
2. $f(\alpha + 1) = f(\alpha) \setminus \bigcup\{B_K \times A_K \mid K \in \mathcal{K}_{f(\alpha)}\}$,
3. $f(\alpha) = \bigcap_{\beta < \alpha} f(\beta)$, if α is a limit ordinal.

Since $\aleph \not\leq |Z|$, f cannot be injective. Thus there exists some $\alpha \in \aleph$ with $f(\alpha + 1) = f(\alpha)$. For this α, $\mathcal{K}_{f(\alpha)}$ must be empty, i.e., $f(\alpha)$ is a linear order relation on X.

The following self–explanatory table summarizes some of the results of this section.

---

[53] [KiWa55]

Table 4.41. [54]

	Filters				Ideals			
	maxim. exist	prime exist	maxim. extens.	prime extens.	maxim. exist	prime exist	maxim. extens.	prime extens.
lattice	**AC**	**False**	**AC**	**False**	**AC**	**False**	**AC**	**False**
distributive	**AC**	**PIT**	**AC**	**PIT**	**AC**	**PIT**	**AC**	**PIT**
Boolean	**PIT**	**PIT**	**PIT**	**PIT**	**PIT**	**PIT**	**PIT**	**PIT**
frame[55]	**PIT**	**PIT**	**PIT**	**PIT**	**AC**	**PIT**	**AC**	**PIT**
open	**PIT**	**True**	**PIT**	**PIT**	**AC**	**True**	**AC**	**PIT**
closed	**AC**	**True**	**AC**	**PIT**	**PIT**	**True**	**PIT**	**PIT**
zero[56]	**True**	**True**	**PIT**	**PIT**	**True**	**True**	?	**PIT**
powerset	**True**	**True**	**PIT**	**PIT**	**True**	**True**	**PIT**	**PIT**

Exercises to Section 4.3:

E 1. Consider the following conditions:
(1) Each non–empty linearly ordered set without a largest element has an increasing sequence.
(2) Each linearly ordered set without a decreasing sequence is well–ordered.
(3) Each infinite linearly ordered set contains an increasing or a decreasing sequence.
(4) Each linearly ordered D–finite set is finite.
Show that (1) $\Leftrightarrow$ (2) $\Rightarrow$ (3) $\Leftrightarrow$ (4).

E 2. Show that **UFT**($\mathbb{N}$) implies that there are precisely $2^{2^{\aleph_0}}$ ultrafilters on $\mathbb{N}$.
[Hint: Observe that the Cantor cube $\mathbf{2}^{\mathbb{R}}$ is a separable Hausdorff space.]

E 3. Show that **WUF**($\mathbb{N}$) guarantees the existence of at least $2^{\aleph_0}$ free ultrafilters on $\mathbb{N}$.
[Hint: Use the fact (see the proof of Theorem 7.21(3)) that there exists a set X of infinite subsets of $\mathbb{N}$ such that $|X| = 2^{\aleph_0}$ and the intersection of any two members of X is finite.]

---

54 Cf. [Her2005]
55 See Exercise E 13.
56 See Exercise E 6.

E 4. Let X be an amorphous set (see Exercises to Section 4.1, E 11). Show that:
(1) There is precisely one free ultrafilter on X.
(2) There are precisely n free ultrafilters on $X \times \{0, 1, \ldots, n-1\}$.
(3) For each Hausdorff topology τ on X, the space (X, τ) has at most one non–isolated point.

E 5. Show the equivalence of the conditions:
(1) Each distributive lattice contains a prime filter.
(2) For distributive lattices each filter is contained in a prime one.
(3) **UFT**.
[Hint: For the implication (3) $\Rightarrow$ (2) proceed as in the proof of Theorem 4.37, (4) $\Rightarrow$ (1).]

E 6. (1) Show that, for a non–empty topological space X the set $\mathfrak{Z}(X)$ of all *zero–sets* (i.e., of all preimages $f^{-1}(0)$ of 0 under some continuous map $f\colon X \to \mathbb{R}$) forms a lattice.
(2) Call a lattice a *zero–lattice* provided it is isomorphic to $\mathfrak{Z}(X)$ for some non–empty topological space X, and show the equivalence of the following conditions:
(a) In a zero–lattice, each filter can be enlarged to a maximal one.
(b) In a zero–lattice, each filter can be enlarged to a prime one.
(c) **UFT**.

E 7. Show that each set X with $|X| \geq 2$ can be lattice–ordered.

E 8. Can each set be ordered as a distributive lattice?

E 9. Show that:
(1) Every finite set can be linearly ordered.
(2) For each finite partially ordered set (F, R) there exists a linear order relation S on F with $R \subseteq S$.

E 10. A complete lattice L is called *completely distributive* iff for every family $(A_i)_{i\in I}$ and every function

$$x\colon \bigcup_{i\in I}(\{i\} \times A_i) \to L$$

the following equation holds:

$$\bigwedge_{i\in I} \bigvee_{a\in A_i} x(i,a) = \bigvee_{(a_i)\in \prod_{i\in I} A_i} \bigwedge_{i\in I} x(i, a_i).$$

Show the equivalence of:
(a) **AC**.
(b) The chain $\mathbf{2} = \{0, 1\}$ is completely distributive.

(c) Every complete chain is completely distributive.
(d) Every powerset–lattice is completely distributive.

E 11. [57] Construct a chain in $\mathcal{P}\mathbb{N}$ that is order–isomorphic to $\mathbb{R}$.
[Hint: Observe that $\mathcal{P}\mathbb{N}$ and $\mathcal{P}\mathbb{Q}$ are order–isomorphic.]

E 12. [58] Let A be a subset of a partially ordered set X. An upper bound s of A is called a *constructive supremum* of A provided that there exists a function $f : X \to A$ such that

$$s \leq x \Leftrightarrow f(x) \leq x \qquad \text{for each } x \in X.$$

Show the equivalence of:
(1) All suprema are constructive.
(2) **AC**.

E 13. A complete lattice L, satisfying the equation

$$x \wedge \sup A = \sup\{x \wedge a \mid a \in A\}$$

for all $x \in L$ and $A \subset L$ is called a *frame*. Show that
a) Each open lattice is a frame.
b) Each frame L is *pseudocomplement*, i.e., for each $x \in L$ the set $\{y \in L \mid x \wedge y = 0\}$ has a largest member.

4.4 Disasters in Algebra I: Vector Spaces

In **ZFC** every vector space is uniquely determined, up to isomorphism, by a single cardinal number, its dimension. Each of the two fundamental results which together enable us to associate dimension with a given vector space fail badly in **ZF**.

Disaster 4.42. [59] The following can happen:

1. Vector spaces may have no bases[60].
2. Vector spaces may have two bases with different cardinalities.

Even stranger phenomena may haunt us:

---

57 [Sie58, p. 78]
58 [Ern2001]
59 [Laeu62/63]
60 Observe, however, that the existence of a basis for every vector space may, besides desirable consequences, also have some rather ugly ones. See Section 5.1.

Disaster 4.43. [61] The following can happen:

1. In a vector space no non–trivial linear subspace need have a complement.
2. A vector space may have only finite–dimensional proper subspaces, but fail to be finite–dimensional itself.

Theorem 4.44. [62] *Equivalent are:*

1. *Every vector space has a basis.*
2. **AC**.

Proof. **(1) $\Rightarrow$ (2)** In view of Theorem 2.4 it suffices to show that (1) implies **AMC**. Let $(X_i)_{i\in I}$ be a family of pairwise disjoint non–empty sets. Consider an arbitrary field k and let $k(X)$ be the field of rational functions in the variables $x \in X = \bigcup\limits_{i\in I} X_i$ over k. For monomials, i.e., elements of $k(X)$ which have the form $p = \alpha \cdot x_1^{n_1} \cdot x_2^{n_2} \cdots x_m^{n_m}$, we define, for each $i \in I$, the i–degree of p as $d_i(p) = \sum\limits_{x_k \in X_i} n_k$. An element of $k(X)$, $\alpha = \frac{p_1+\cdots+p_n}{q_1+\cdots+q_m}$ where the p_k and q_k are monomials, will be called i–homogeneous of degree d provided that all q_k have the same i–degree, say d_1, and all p_k have the same i–degree $d_2 = d_1 + d$. Then $K = \{a \in k(X) \mid a \text{ is } i\text{–homogeneous of degree } 0 \text{ for each } i \in I\}$ is a subfield of $k(X)$. Thus $k(X)$ is a vector space over K. By (1), $k(X)$ has a basis B. For each $x \in X$ the monomial x can be expressed uniquely in the form $x = \sum\limits_{b\in B(x)} a_b(x) \cdot b$, where $B(x)$ is a finite subset of B and each $a_b(x) \in K \setminus \{0\}$. Let x and y be elements of the same X_i. Then

$$y = \frac{y}{x} \cdot x = \sum_{b\in B(x)} \frac{y}{x} \cdot a_b(x) \cdot b = \sum_{b\in B(y)} a_b(y) \cdot b.$$

Since $\frac{y}{x} \in K$, this implies $B(x) = B(y)$ and $\frac{a_b(y)}{y} = \frac{a_b(x)}{x}$ for each $b \in B(x)$. Thus the sets $B(x)$ and the elements $\frac{a_b(x)}{x}$ depend only on i, and not on the particular x in X_i. Let us call them B_i resp. $\alpha(b,i)$. Since the $a_b(x)$ are i–homogeneous of degree 0, the $\alpha(b,i) = \frac{a_b(x)}{x}$ are i–homogeneous of degree -1. Thus, if $\alpha(b,i)$ is written as a quotient of polynomials in reduced form, some $x \in X_i$ must occur in the denominator. Hence the set F_i, consisting of all $x \in X_i$ that occur in the denominator of $\alpha(b,i)$ in its reduced form for some $b \in B_i$, is a non–empty, finite subset of X_i. This establishes **AMC**.

(2) $\Rightarrow$ (1) is well known.

Theorem 4.45. [63] *For each field k, the following are equivalent:*

1. *Every subspace S of a vector space V over k has a linear complement S' (i.e., $S \cap S' = \{0\}$ and $S + S' = V$).*
2. **AC**.

---

[61] [Laeu62/63]. See also Exercise E 1.

[62] [Blass84]

[63] [Blei64]

Proof. **(1) ⇒ (2)** In view of Theorem 2.4 it suffices to show that (1) implies **AMC**. Let $(X_i)_{i\in I}$ be a family of pairwise disjoint non–empty sets. For each $i \in I$ let $V_i = k^{(X_i)}$ be the direct sum of X_i copies of k, with the canonical base $(e_x)_{x\in X_i}$. Let S_i be the linear subspace of V_i, consisting of all $v = \sum_{x\in X_i} \alpha_x e_x$ in V_i with $\sum_{x\in X_i} \alpha_x = 0$.

Consider the direct sum $S = \bigoplus_{i\in I} S_i$ as a linear subspace of the direct sum $V = \bigoplus_{i\in I} V_i$. By (1), there exists a linear subspace S' of V with $S \cap S' = \{0\}$ and $S + S' = V$ For any $x \in X_i$ consider the element e_x from the canonical base of V_i as an element of V. Then there exist unique elements $s(x)$ in S and $s'(x)$ in S' with $e_x = s(x) + s'(x)$. If x and y belong to the same X_i, then $s'(x) - s'(y) = (e_x - s(x)) - (e_y - s(y)) = (e_x - e_y) + (s(y) - s(x))$ belongs to S' and to S, since $(e_x - e_y) \in S_i$. Therefore $s'(x) = s'(y)$. This element depends only on i and not on the particular x in X_i. Let us call it s'_i.

Let $s'_i = \sum_{j\in I} v_j$ with $v_j \in V_j$, and $v_i = \sum_{x\in X_i} \alpha_x e_x$ with $\alpha_x \in k$ be the canonical expressions. Then $F_i = \{x \in X_i \mid \alpha_x \neq 0\}$ is finite. Since for $x \in X_i$, $s(x) = e_x - s'_i = e_x - \sum_{j\in I} v_j = (e_x - v_i) - \sum_{j\neq i} v_j \in S$, we conclude $(e_x - v_i) \in S_i$, hence $1 - \sum_{x\in X_i} \alpha_x = 0$, hence $F_i \neq \emptyset$. This establishes **AMC**.

(2) ⇒ (1) is well known.

We may ask whether, like in Theorem 4.45, we can restrict attention in Theorem 4.44 to vector spaces over $\mathbb{R}$ resp. $\mathbb{Q}$. The answer to these questions is unknown. However, there is a slightly weaker result in the case of $\mathbb{Q}$.

Lemma 4.46. *Let k be a field. If for every vector space V over k every generating set contains a basis, then for each family $(V_i)_{i\in I}$ of vector spaces over k and each family $(G_i)_{i\in I}$ of generating sets G_i of V_i there exists a family $(B_i)_{i\in I}$ of bases B_i of V_i with $B_i \subseteq G_i$ for each $i \in I$.*

Proof. Let $V = \bigoplus_{i\in I} V_i$ be the direct sum of the V_i's, with canonical inclusion maps $j_i \colon V_i \to V$. Then $G = \bigcup_{i\in I} j_i[G_i]$ is a generating set for V. Let $B \subseteq G$ be a basis for V. Then $B_i = \{x \in V_i \mid j_i(x) \in B\}$ is a basis for V_i with $B_i \subseteq G_i$ for each $i \in I$.

Theorem 4.47. [64] *Equivalent are:*

1. *In every vector space over $\mathbb{Q}$ each generating set contains a basis.*
2. **AC**.

[64] [Ker98]

Proof. **(1) ⇒ (2)** Again by Theorem 2.4 it suffices to show that (1) implies **AMC**. Let $(X_i)_{i\in I}$ be a family of non–empty sets. Assume w.l.o.g. that each X_i contains at least two elements. For each $i \in I$, let V_i be the linear subspace of the product space $\mathbb{Q}^{X_i}$, consisting of all $(\alpha_x)_{x\in X_i}$ that are almost constant, i.e., for which there exists a finite subset F of X_i such that $(\alpha_x)_{x\in(X_i\setminus F)}$ is a constant family. Each V_i is generated by the set G_i, consisting of all elements $v = (\alpha_x)_{x\in X_i}$ of V_i such that there exists exactly one element $x = x(v)$ in X_i such that $(\alpha_y)_{y\in(X_i\setminus\{x\})}$ is a constant family.

By (1) and Lemma 4.46 there exists a family $(B_i)_{i\in I}$ of bases B_i of V_i with $B_i \subseteq G_i$. For each $i \in I$ consider the element $a_i = (\alpha_x)_{x\in X_i}$ with each $\alpha_x = 1$.

Then there exists a unique non–empty, finite subset B_i' of B_i such that a_i is expressible as a linear combination

$$a_i = \sum_{b\in B_i'} \alpha_b b \text{ with each } \alpha_b \in (\mathbb{Q}\setminus\{0\}).$$

Thus $F_i = \{x(b) \mid b \in B_i'\}$ is a non–empty finite subset of X_i. This establishes **AMC**.

(2) ⇒ (1) is well known.

Returning to statement (2) of Disaster 4.42, no choice principle is known that is equivalent to the statement that any two bases of a vector space have the same cardinality. However, the latter fact follows already from **PIT**[65].

In **ZFC** every vector space is *injective*[66] and *projective*[67]. Since projectivity of V follows from V being *free* (i.e., V has a basis), Theorem 4.44 casts doubt on the idea that in **ZF** every vector space is projective. The situation is even worse:

Disaster 4.48. 1. Vector spaces may fail to be injective.
2. Vector spaces, even free ones, may fail to be projective.

Theorem 4.49. *For each field k, the following are equivalent:*

1. *Every vector space over k is injective.*
2. *Every vector space over k is projective.*
3. *Every free vector space over k is projective.*
4. **AC**.

---

[65] [Hal66]

[66] V is called *injective* iff every linear map $f: U \to V$ from a linear subspace U of a vector space W can be extended to a linear map $\bar{f}: W \to V$.

[67] V is called *projective* iff for every linear map $f: V \to U$ and every linear surjection $g: W \to U$ there exists a linear map $\bar{f}: V \to W$ with $f = g \circ \bar{f}$.

Proof. Since clearly **(4)** $\Rightarrow$ **(1)**, **(4)** $\Rightarrow$ **(2)**, and **(2)** $\Rightarrow$ **(3)** it suffices to show that
(1) $\Rightarrow$ (4) and (3) $\Rightarrow$ (4).

(1) $\Rightarrow$ **(4)** By Theorem 4.45 it suffices to show that every subspace A of a vector space V has a linear complement. Since A, by (1), is injective there exists a k–linear map $f\colon V \to A$ such that the diagram

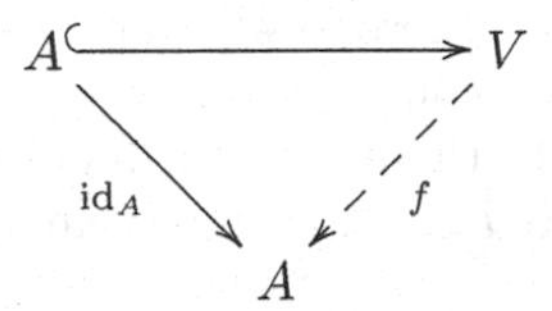

commutes. Then $B = \{x \in V \mid f(x) = 0\} = \{x - f(x) \mid x \in V\}$ is a subspace of V with $A \cap B = \{0\}$ and $A + B = V$, since $x = f(x) + (x - f(x))$. Thus B is a linear complement of A in V.

(3) $\Rightarrow$ **(4)** Let $(X_i)_{i\in I}$ be a family of non–empty sets. Let $X = \{(i,x) \mid i \in I \text{ and } x \in X_i\}$ be the disjoint union of the X_i's. Let $F(I) = k^{(I)}$ resp. $F(X) = k^{(X)}$ be the direct sums of I resp. X copies of k (i.e., the canonical free k–vector spaces over I resp. X). Then the map $f\colon X \to I$, defined by $f(i,x) = i$, induces a linear map $\bar{f}\colon F(X) \to F(I)$. Since f is surjective, so is $\bar{f}$. Thus, by projectivity of the free vector space $F(I)$, there exists a linear map $g\colon F(I) \to F(X)$ such that the diagram

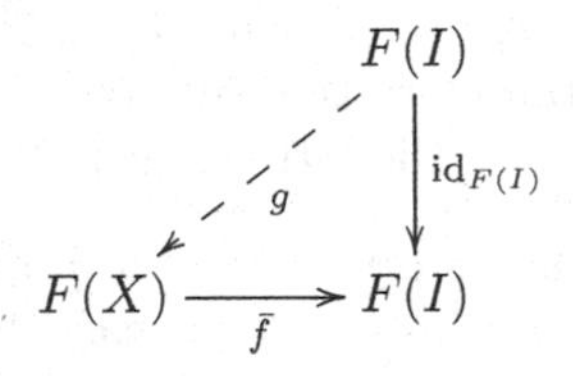

commutes. For each $i \in I$ consider $e_i = (\delta_{ij})_{j\in I}$. Then $g(e_i) = (k(j,x))_{(j,x)\in X}$ and $e_i = \bar{f}(g(i)) = \left(\sum_{x\in X_j} k(j,x)\right)_{j\in I}$. The set $F_i = \{x \in X_i \mid k(i,x) \neq 0\}$ is, by definition of $k^{(X)}$, finite. Moreover, the equation

$$1 = \delta_{ii} = \sum_{x\in X_i} k(i,x) \quad \text{implies that} \quad F_i \neq \emptyset.$$

Thus $(F_i)_{i\in I}$ is a family of non–empty, finite subsets F_i of X_i. Consequently **AMC**, holds, and so does **AC** via Theorem 2.4.

Exercises to Section 4.4:

E 1. [68] Let p be a prime and let X be a vector space over $\mathbb{Z}_p$ with amorphous[69] underlying set[70]. Show that:

---

[68] [Hic76, III.1]

[69] See Exercises to Section 4.1, E 11.

[70] Such X exist; see [Hic76, II.2].

(1) Each finitely generated linear subspace of X is finite.
(2) Every proper linear subspace of X is finite.
(3) X has no basis.

4.5 Disasters in Algebra II: Categories

Category theory heavily depends on choice principles. In **ZF**, already the basic Adjoint Functor Theorems fail dramatically:

Disaster 4.50. [71]

1. The Adjoint Functor Theorem fails, i.e., there exists a limit–preserving functor $G\colon \mathbf{A} \to \mathbf{B}$ with complete domain which satisfies the solution–set–condition, but has no coadjoint.
2. The Special Adjoint Functor Theorem fails, i.e., there exists a strongly complete category $\mathbf{A}$ with a coseparator and a functor $G\colon \mathbf{A} \to \mathbf{B}$ that preserves strong limits, but has no coadjoint.
3. There exists a functor $G\colon \mathbf{A} \to \mathbf{B}$ such that each $\mathbf{B}$–object has a G–universal arrow, carried by a $\mathbf{B}$–identity, but which has no coadjoint, not even a right inverse.

How can these disasters be prevented? Only by the Axiom of Choice:

Theorem 4.51. [72] *Equivalent are:*

1. *The Adjoint Functor Theorem holds.*
2. *The Special Adjoint Functor Theorem holds.*
3. *Every functor $G\colon \mathbf{A} \to \mathbf{B}$, such that each $\mathbf{B}$-object has a G-universal arrow, has a coadjoint.*
4. **AC** *for classes.*

Proof. That (**4**) implies (**1**), (**2**), and (**3**) is well known. See, e.g., [AHS2004, 18.12, 18.17, and 19.1]

To show that each of the conditions (**1**), (**2**), and (**3**) imply (**4**), consider a family $(X_i)_{i\in I}$ of non–empty sets, indexed by some class I. Construct a functor $G\colon \mathbf{A} \to \mathbf{B}$ as follows:

$\mathbf{A}$ and $\mathbf{B}$ are the categories naturally associated with the preordered classes $(A, \leq)$ and $(B, \leq)$, where

$B = I \uplus \{0, 1\}$ is obtained from the discretely ordered class I by adding a first element 0 and a last element 1,

$A = \{(x, i) \mid i \in I \text{ and } x \in X_i\} \uplus \{0, 1\}$ is preordered by having 0 as first element, 1 as last element and

$$(x, i) \leq (y, j) \text{ iff } i = j.$$

[71] [Den2003]
[72] [Den2003]

$G\colon \mathbf{A} \to \mathbf{B}$ is defined by $G(0) = 0$, $G(1) = 1$, and $G(x, i) = i$. Then the premisses of each of the conditions (**1**), (**2**), (**3**) are satisfied. In particular the G–universal arrows for B–objects have the form $\mathrm{id}_i\colon i \to G(x, i)$ with $x \in X_i$. Thus there exists a coadjoint $F\colon \mathbf{B} \to \mathbf{A}$ for G. For each $\mathbf{B}$–object i, $F(i) = (x_i, i)$ for a unique element x_i of X_i. Thus $(x_i) \in \prod_{i\in I} X_i$.

Exercises to Section 4.5:

E 1. Show that **AC** holds iff every epimorphism in **Set** is a retraction.

E 2. [73] Show that for a set I the following conditions are equivalent:
(1) I is projective.
(2) Every epimorphism with codomain I is a retraction.
(3) $\prod_{i\in I} X_i \neq \emptyset$ for every family $(X_i)_{i\in I}$ of non–empty sets.

4.6 Disasters in Elementary Analysis: The Reals and Continuity

Elementary analysis in **ZF** suffers from various defects. Before we analyze these, let us first point out several basic properties of the reals and of real functions in **ZFC** that remain valid in the **ZF**–setting:

Theorem 4.52. *1. $\mathbb{R}$ and all its subspaces are metrizable, hence normal.*
2. $\mathbb{R}$ and all its subspaces are second countable, *i.e., have at most countable bases.*
3. $\mathbb{R}$ is separable, *i.e., $\mathbb{R}$ has an at most countable, dense subset.*
4. A subspace of $\mathbb{R}$ is connected iff it is an interval, *i.e., contains with any elements x and y each element between x and y.*
5. A subspace of $\mathbb{R}$ is compact iff it is bounded and closed in $\mathbb{R}$.
6. For each bounded, infinite subset of $\mathbb{R}$ there exists an accumulation point in $\mathbb{R}$.
7. $\mathbb{R}$ is σ–compact, i.e., a countable union of compact subspaces.
8. A function $f\colon \mathbb{R} \to \mathbb{R}$ is continuous iff it is sequentially continuous.
9. A function $f\colon [0, 1] \to \mathbb{R}$ is continuous iff it is uniformly continuous.

Proof. In most cases the **ZFC**–proof carries over to the **ZF**–setting. However, see Proposition 3.30 for (5), Theorem 3.15 for (8), and Proposition 3.14 for (9).

Unfortunately, however, the close and useful ties that exist in **ZFC** between static (ϵ–δ–definitions) and dynamic (use of sequences) aspects break

[73] Cf. Exercises to Section 2.1, E 4.

apart in the **ZF**–setting, where the construction of sequences — resp. more generally: of countable sets — with specified properties in most cases relies heavily on the condition **CC**($\mathbb{R}$). In fact, **CC**($\mathbb{R}$) turns out to be equivalent to a surprising number of familiar statements in elementary analysis.

Disaster 4.53. [74] The following can happen:

1. $\mathbb{R}$ may fail to be *Fréchet*, i.e., not every accumulation point x of a subset A may be reachable[75] by a sequence (a_n) in A.
2. $\mathbb{R}$ may fail to be *sequential*, i.e., there may be non–closed, sequentially closed[76] subsets of $\mathbb{R}$.
3. $\mathbb{R}$ may fail to be Lindelöf.
4. All Lindelöf spaces of $\mathbb{R}$ may be compact.
5. Subspaces of $\mathbb{R}$ may fail to be separable.
6. Complete subspaces of $\mathbb{R}$ may fail to be closed in $\mathbb{R}$.
7. Sequentially compact subspaces of $\mathbb{R}$ may fail to be bounded or to be closed in $\mathbb{R}$.
8. [77] Functions $f\colon \mathbb{R} \to \mathbb{R}$ may be sequentially continuous at some point x, but fail to be continuous at x.
9. [78] Functions $f\colon X \to \mathbb{R}$, defined on some subspace X of $\mathbb{R}$, may be sequentially continuous, but fail to be continuous.
10. Infinite subsets of $\mathbb{R}$ may be D–finite.

Proof. First, consider a model[79] of **ZF** with an infinite D–finite subset X of $\mathbb{R}$. Assume, without loss of generality, that X is bounded. By Theorem 4.52(6) there exists an accumulation point a of X in $\mathbb{R}$. Assume further, without loss of generality, that a is not contained in X. Then X is sequentially closed, but not closed in $\mathbb{R}$. Thus (1) and (2) may occur. Furthermore, the subspace X of $\mathbb{R}$ is complete and sequentially compact, but fails to be separable or to be closed in $\mathbb{R}$. Thus (5), (6), and (7) may occur. The function $f\colon \mathbb{R} \to \mathbb{R}$, defined by $f(x) = \begin{cases} 1, \text{ if } x \in X \\ 0, \text{ otherwise} \end{cases}$, is sequentially continuous at a but fails to be continuous at a. Its restriction to $A = X \cup \{a\}$ is sequentially continuous, but not continuous. Thus (8) and (9) may occur.

That (3) may occur, follows from Theorem 3.8, since with $\mathbb{R}$ also its closed subspace $\mathbb{N}$ would be Lindelöf. Finally, the possible occurrence of (4) will be shown in Section 7.1. (See Theorem 7.2.)

How much choice is needed to eliminate the above disasters?

---

[74] [Jae65], [Jec68], [Bru82], [Her2002], [Gut2003].
[75] i.e., $(a_n) \to x$.
[76] A is *sequentially closed* in $\mathbb{R}$ iff no point outside A is reachable by a sequence in A.
[77] Cf. this disaster with Theorem 3.15.
[78] Cf. this disaster with Theorem 3.15.
[79] Such models exist, e.g., Cohen's First Model A4 (M1 in [HoRu98]).

Theorem 4.54. [80] *Equivalent are:*

1. *$\mathbb{R}$ is Fréchet.*
2. *Each subspace of $\mathbb{R}$ is sequential.*
3. *$\mathbb{R}$ is Lindelöf.*
4. *Each subspace of $\mathbb{R}$ is Lindelöf.*
5. *Each second countable topological space is Lindelöf.*
6. *Each subspace of $\mathbb{R}$ is separable.*
7. *Each second countable topological space is separable.*
8. *A function $f\colon \mathbb{R} \to \mathbb{R}$ is continuous at some point x iff it is sequentially continuous at x.*
9. *A function $f\colon X \to \mathbb{R}$, defined on some subspace X of $\mathbb{R}$, is continuous iff it is sequentially continuous.*
10. $\mathbf{CC}(\mathbb{R})$.

Proof. **(1)** $\Rightarrow$ **(2)** With $\mathbb{R}$ each subspace of $\mathbb{R}$ is Fréchet, and thus sequential.

(2) $\Rightarrow$ **(9)** If $f\colon X \to \mathbb{R}$ is sequentially continuous then the f–preimage of each closed set in $\mathbb{R}$ is sequentially closed in X, thus, by (2), closed in X. Consequently f is continuous. The inverse implication holds trivially in **ZF**.

(9) $\Rightarrow$ **(10)** By Theorem 3.8 it suffices to show that under (9) every unbounded subset A of $\mathbb{R}$ contains an unbounded sequence. Let $h\colon \mathbb{R} \to (0,1)$ be a homeomorphism. Without loss of generality, 0 is an accumulation point of $h[A]$. Define $X = h[A] \cup \{0\}$ and $f\colon X \to \mathbb{R}$ by $f(x) = \begin{cases} 0, \text{ if } x \in h[A] \\ 1, \text{ if } x = 0 \end{cases}$. Then f is not continuous, thus, by (9), not sequentially continuous. Thus there exists a sequence (b_n) in $h[A]$ that converges to 0. Consequently $(h^{-1}(b_n))$ is an unbounded sequence in A.

(10) $\Rightarrow$ **(5)** Since $\mathbb{R}$ and $\mathcal{P}(\mathbb{N})$ have the same cardinal number $2^{\aleph_0}$, $\mathbf{CC}(\mathbb{R})$ implies $\mathbf{CC}(\mathcal{P}(\mathbb{N}))$, i.e., countable products $\prod_{m\in M} U_m$ of non–empty subset U_m of $\mathcal{P}(\mathbb{N})$ are non–empty. Let X be a second countable topological space with a basis $\mathfrak{B} = \{B_n \mid n \in \mathbb{N}\}$, and let $\mathcal{C}$ be an open cover of X. For each $n \in \mathbb{N}$, define $U_n = \{C \in \mathcal{C} \mid B_n \subseteq C\}$. Consider $M = \{m \in \mathbb{N} \mid U_m \neq \emptyset\}$. Then there exists an element $(C_m)_{m\in M}$ in $\prod_{m\in M} U_m$. Consequently $\{C_m \mid m \in M\}$ is an at most countable subcover of $\mathcal{C}$.

(5) $\Rightarrow$ **(4)** $\Rightarrow$ **(3)** Immediate.

(3) $\Rightarrow$ **(10)** By (3), $\mathbb{N}$ as a closed subspace of $\mathbb{R}$ must be Lindelöf. Thus, by Theorem 3.8, (10) holds.

(10) $\Rightarrow$ **(7)** $\Rightarrow$ **(6)** Immediate.

(6) $\Rightarrow$ **(1)** Let a be an accumulation point of some subset X of $\mathbb{R}$. X, being separable, contains a countable dense subset C. Consequently a is an accumulation point of C. Thus C, being countable, contains a sequence converging to a.

---

[80] [HeSt97]

(10) ⇒ **(8)** Immediate.

(8) ⇒ **(1)** Let a be an accumulation point of some subset X of $\mathbb{R}$. If no sequence in X converges to a, then $a \notin X$ and the function $f: \mathbb{R} \to \mathbb{R}$, defined by $f(x) = \begin{cases} 1, \text{ if } x \in X \\ 0, \text{ otherwise} \end{cases}$, is sequentially continuous at a, but not continuous at a. This contradicts condition (8).

Theorem 4.55. [81] *Equivalent are:*

1. $\mathbb{R}$ *is sequential,*
2. *Complete = closed in* $\mathbb{R}$*, for subspaces of* $\mathbb{R}$.
3. *Compact = complete and bounded, for subspaces of* $\mathbb{R}$.
4. *Compact = sequentially compact, for subspaces of* $\mathbb{R}$.
5. *Complete subspaces of* $\mathbb{R}$ *are separable.*
6. *Complete, unbounded subspaces of* $\mathbb{R}$ *contain unbounded sequences.*
7. $\mathbf{CC}(c\mathbb{R})$*, i.e.,* $\prod_{n\in\mathbb{N}} X_n \neq \emptyset$ *for every sequence of non–empty, complete subspaces* X_n *of* $\mathbb{R}$.
8. $\mathbf{AC}(c\mathbb{R})$*, i.e.,* $\prod_{i\in I} X_i \neq \emptyset$ *for every family of non–empty, complete subspaces* X_i *of* $\mathbb{R}$.

Proof. **(1)** ⇒ **(2)** Immediate, since every complete subspace of $\mathbb{R}$ is sequentially closed in $\mathbb{R}$.

(2) ⇒ **(8)** Immediate in view of Exercises to Section 1.1, E 2(5).

(8) ⇒ **(7)** Obvious.

(7) ⇒ **(5)** Let X be a complete subspace of $\mathbb{R}$. Consider the set M of all pairs $(p,q) \in \mathbb{Q}^2$ for which the set $C_{(p,q)} = [p,q] \cap X$ is not empty. By (7), there exists an element $(x_m)_{m\in M} \in \prod_{m\in M} C_m$. Consequently $\{x_m \mid m \in M\}$ is an at most countable, dense subset of X.

(5) ⇒ **(6)** Immediate.

(6) ⇒ **(1)** Let X be sequentially closed in $\mathbb{R}$. Assume that there exists an accumulation point of X in $\mathbb{R}$ with $a \notin X$. Without loss of generality we may assume that $X \subseteq (0,1)$ and $a = 1$. Let $h: (0,1) \to \mathbb{R}$ be an order–preserving homeomorphism with $|x-y| \leq |h(x) - h(y)|$ for each pair (x,y) in $(0,1)^2$. Then $h[X]$ is complete and unbounded. Thus, by (6), there exists an unbounded sequence (y_n) in $h[X]$. Without loss of generality we may assume that (y_n) is monotone increasing. Thus the sequence $(h^{-1}(y_n))$ converges to 1, contradicting the stipulation that X is sequentially closed in $\mathbb{R}$.

(2) ⇒ **(3)** Immediate in view of Theorem 4.52(5).

(3) ⇒ **(2)** Let X be a complete subspace of $\mathbb{R}$. Then, for each $n \in \mathbb{N}$, the space $X \cap [-n,n]$ is complete and bounded, thus by (3) compact, thus closed in $\mathbb{R}$. Consequently, X is closed in $\mathbb{R}$.

(3) ⇒ **(4)** Let X be a sequentially compact subspace of $\mathbb{R}$. Then X is complete. Thus, by (3), it suffices to show that X is bounded. Assume that X

---
[81] [Gut2003]

is unbounded, then by (6) — which is implied by (3), as shown above — X contains an unbounded sequence, and consequently a sequence without convergent subsequence. This contradicts the stipulation that X is sequentially compact.

(4) $\Rightarrow$ **(3)** Immediate since each complete and bounded subspace of $\mathbb{R}$ is sequentially compact.

If $\mathbb{R}$ is Fréchet, then $\mathbb{R}$ is sequential. Does the converse implication hold? The answer is no, as we will see below.

Definition 4.56. [82] ω–**CC**$(\mathbb{R})$ *states that for every sequence* $(X_n)_{n\in\mathbb{N}}$ *of non–empty subsets of* $\mathbb{R}$, *there exists a sequence* $(C_n)_{n\in\mathbb{N}}$ *of non–empty, at most countable subsets* C_n *of* X_n.

Proposition 4.57. [83] *Each of the following statements implies the succeeding ones:*

1. $\mathbb{R}$ *is the countable union of countable sets.*
2. ω–**CC**$(\mathbb{R})$.
3. $\mathbb{R}$ *is sequential.*
4. **Fin**$(\mathbb{R})$.

Proof. **(1)** $\Rightarrow$ **(2)** Let $(X_n)_{n\in\mathbb{N}}$ be a sequence of non–empty subsets of $\mathbb{R}$ and let $\mathbb{R}$ be the union of the sequence $(A_m)_{m\in\mathbb{N}}$ of countable sets A_m. For each $n \in \mathbb{N}$, define

$$m(n) = \min\{m \in \mathbb{N} \mid (X_n \cap A_m \neq \emptyset\}$$

Then, for each $n \in \mathbb{N}$, $C_n = X_n \cap A_{m(n)}$ is a non–empty at most countable subset of X_n.

(2) $\Rightarrow$ **(3)** Let a be an accumulation point of some sequentially closed subset X of $\mathbb{R}$. Then, for each $n \in \mathbb{N}$, the set

$$X_n = X \cap [a - \frac{1}{n+1},\ a + \frac{1}{n+1}]$$

is non–empty. Thus, by (2), there exists a sequence $(C_n)_{n\in\mathbb{N}}$ of non–empty, at most countable subsets C_n of X_n. Since X is sequentially closed, each $x_n = \inf C_n$ belongs to X, and thus a, the limit of (x_n), belongs to X as well. Consequently X is closed in $\mathbb{R}$.

(3) $\Rightarrow$ **(4)** Immediate, since all infinite, D–finite subsets of $\mathbb{R}$ are sequentially closed but not closed in $\mathbb{R}$.

Thus we get the following diagram:

[82] [KeTa2001], [Gut2003].

[83] [Gut2003]

Diagram 4.58.

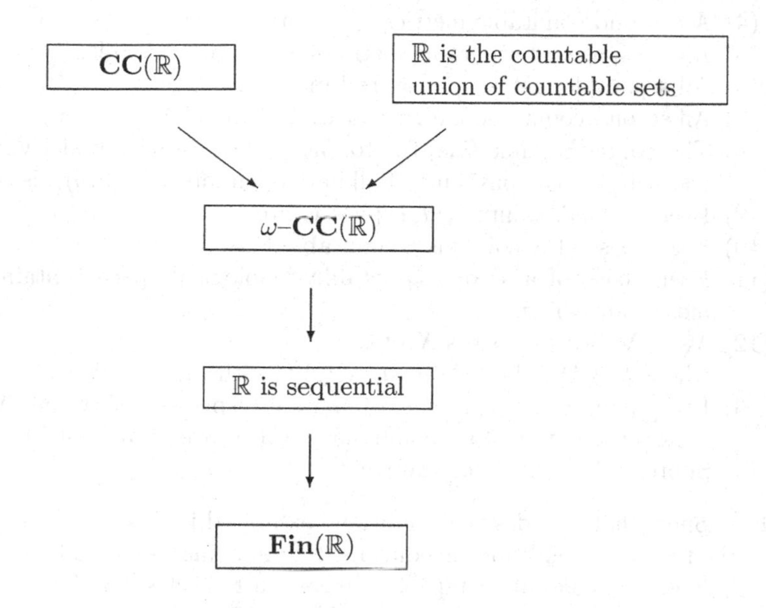

Remark 4.59. [84] There are models[85] of **ZF**, in which $\mathbb{R}$ is the countable union of countable sets[86] but $\mathbf{CC}(\mathbb{R})$ fails. Thus in these models $\mathbb{R}$ is sequential[87], but not Fréchet. It is not known whether $\mathbf{Fin}(\mathbb{R})$ implies that $\mathbb{R}$ is sequential.

Exercises to Section 4.6:

E 1. Show that:
 (1) Every open subspace of $\mathbb{R}$ is separable.
 (2) Every closed subspace of $\mathbb{R}$ is separable.
 (3) Ever Lindelöf subspace of $\mathbb{R}$ is separable.

E 2. Every infinite, closed subset of $\mathbb{R}$ is D–infinite.

E 3. [88] Show that, besides the conditions exhibited in Theorems 3.8 and 4.54, each of the following conditions is equivalent to $\mathbf{CC}(\mathbb{R})$:

---

[84] [Gut2003]
[85] E.g., the Feferman–Levy Model A8 (M9 in [HoRu98]).
[86] I.e., the negation of Form 38 in [HoRu98].
[87] Form 74 in [HoRu98].
[88] [BeHe98], [Ker98], [Gut2004], [GiHe2004].

(1) Each subspace of the Cantor Discontinuum is separable.
(2) Each subspace of the space of irrational numbers is separable.
(3) Each subspace of $\mathbb{S}^{\mathbb{N}}$ is separable (where $\mathbb{S}$ is the Sierpiński space).
(4) All second countable metric spaces are separable.
(5) All subspaces of separable metric spaces are separable.
(6) All separable metric spaces are Lindelöf.
(7) All second countable metric spaces are Lindelöf.
(8) The Sorgenfrey line (i.e., the topological space with underlying set $\mathbb{R}$ and with a base consisting of all half–open intervals $[a, b)$), is Lindelöf.
(9) Every second countable T_0–space is Fréchet.
(10) Every base of $\mathbb{R}$ contains a countable base.
(11) Every base of a second countable topological space contains an at most countable base.
(12) $\lambda_X^2 = \lambda_X$ for subspaces X of $\mathbb{R}$, where $\lambda_X(A) = \{x \in X \mid \exists (a_n) \in A^{\mathbb{N}} \text{ with } (a_n) \to x\}$.
(13) Every subset of $\mathbb{R}$ contains a maximal dispersed set (where X is dispersed if any two of its points have a distance of at least 1).
(14) Suprema in $\mathbb{R}$ are constructive[89].

E 4. [90] Show that, besides the conditions exhibited in Theorem 4.55, each of the following conditions is equivalent to $\mathbb{R}$ being sequential:
(1) Every sequentially compact subspace of $\mathbb{R}$ is closed in $\mathbb{R}$.
(2) Every sequentially compact subspace of $\mathbb{R}$ is bounded.
(3) Every sequentially compact subspace of $\mathbb{R}$ is Lindelöf.
(4) Every sequentially compact subspace of $\mathbb{R}$ is separable.
(5) Every complete subspace of $\mathbb{R}$ is separable.
(6) In every complete subspace of $\mathbb{R}$ each Cauchy filter converges.
(7) No proper dense subspace of $\mathbb{R}$ is complete.

E 5. [91] Show that the following conditions are equivalent.
(1) **Fin**($\mathbb{R}$).
(2) For every sequence $(X_n)_{n\in\mathbb{N}}$ of non–empty, D–finite subsets of $\mathbb{R}$, the product $\prod_{n\in\mathbb{N}} X_n$ is non–empty.
(3) For every family $(X_i)_{i\in I}$ of non–empty, D–finite subsets of $\mathbb{R}$, the product $\prod_{i\in I} X_i$ is non–empty.
(4) Every bounded infinite subset of $\mathbb{R}$ contains a convergent injective sequence.
(5) Every D–finite subset of $\mathbb{R}$ is bounded.
(6) $\mathbb{R}$ has no dense D–finite subset.

89 [Ern2001]. Cf. Exercises to Section 4.3, E 12.
90 [Gut2003], [Gut2004].
91 [Bru82], [Gut2004].

4.7 Disasters in Topology I: Countable Sums

Our result vividly demonstrates the horrors of topology without **AC**.

E.K. van Douwen[92]

Even the most innocent of topological questions may be undecidable from the Zermelo–Fraenkel axioms alone.

Good, Tree, and Watson[93]

The formation of countable sums is one of the simplest constructions in topology. In **ZFC** it preserves most of the familiar properties of topological spaces, in particular:

1. metrizability,
2. normality,
3. separability,
4. second countability,
5. the Lindelöf property,
6. dimension zero.

However in **ZF** — though countable sums of metric spaces are still metrizable (see Exercise E 1) — none of the above 6 properties is necessarily preserved under the formation of countable sums. Let us stress in particular that those mathematicians, who have the dangerous habit of not distinguishing between the notions of *metric space* and *metrizable space*, live in an inconsistent world, where countable sums of such spaces are metrizable (Exercise E 1) and at the same time are not necessarily metrizable (Disaster 4.60).

E.K. van Douwen [vDou85] has constructed a model[94] of **ZF** which fails to have the following property:

CC($\mathbb{Z}$): For each sequence $((X_n, \leq_n))_{n\in\mathbb{N}}$ of ordered sets, each of which is order–isomorphic to the set $\mathbb{Z}$ of integers with its natural order, we have $\prod\limits_n X_n \neq \emptyset$.

Disaster 4.60. [95] If **CC**($\mathbb{Z}$) fails there exists a sequence of separable, metrizable, compact spaces (Y_n) with $\dim Y_n = 0$, such that $\sum\limits_n Y_n$ is neither

92 [vDou85]
93 [GoTrWa98]
94 Called (N2(LO))in [HoRu98]. See also [HaMo90].
95 [vDou85]

metrizable, nor normal, nor separable, nor second countable, nor Lindelöf, nor with dimension 0.

Proof. Let (X_n) be a sequence of ordered sets, each order isomorphic to the integers with their natural order, such that $\prod X_n = \emptyset$. For each n, obtain a set Y_n by adding to X_n a first element a_n and a last element b_n, and supply Y_n with the corresponding order–topology. Then the Y_n are all order–isomorphic and thus homeomorphic to the subspace of $\mathbb{R}$ determined by the set

$$S = \{0\} \cup \left\{\frac{1}{n} \mid n \in \mathbb{N}^+\right\} \cup \left\{2 - \frac{1}{n} \mid n \in \mathbb{N}^+\right\} \cup \{2\}.$$

Thus each Y_n is a separable, metrizable, compact space with dimension 0, hence also normal, second countable, and Lindelöf.

Next, consider the sum $Y = \sum_n Y_n$ of the Y_n's. Assume, for simplicity that the Y_n's are pairwise disjoint so that the underlying set of Y is just the union of the Y_n's.

Claim 1: Y is not normal, hence is neither metrizable nor of dimension 0.

Proof of Claim 1: The sets $A = \{a_n \mid n \in \mathbb{N}\}$ and $B = \{b_n \mid n \in \mathbb{N}\}$ are disjoint closed subsets of Y. If U and V would be disjoint neighborhoods of A and B in Y, then, for each $n \in \mathbb{N}$, the set $U \cap X_n$ would be a non–empty, upper–bounded subset of X_n, and would thus contain a largest element x_n. Hence (x_n) would be an element of $\prod_n X_n$, contrary to the assumption.

Claim 2: Y is neither separable nor second countable.

Proof of Claim 2: Since every dense subset of Y contains every x in $\bigcup_n X_n$, and since every base of Y contains each set $\{x\}$ with $x \in \bigcup_n X_n$, separability resp. second countability of Y would imply that $\bigcup_n X_n$ is countable which in turn would imply $\prod_n X_n \neq \emptyset$.

Claim 3: Y is not Lindelöf.

Proof of Claim 3: Consider the open cover

$$\mathfrak{U} = \{Y_n \setminus \{x\} \mid n \in \mathbb{N},\ x \in X_n\}$$

of Y. If $\mathfrak{U}$ would have a countable subcover

$$\mathfrak{V} = \{Y_{n(m)} \setminus \{x_m\} \mid m \in \mathbb{N}\},$$

then, for each $n \in \mathbb{N}$, there would exist a smallest $m = m(n)$ with $n(m) = n$. Thus $(x_{m(n)})_{n \in \mathbb{N}}$ would be an element of $\prod_n X_n$, contrary to the assumption.

Observe that the spaces Y_n, entering in the above construction, are pairwise isomorphic, but not identical. What happens if we form the sum of countably many copies of a fixed space Z, equivalently: the product of Z with a countable discrete space $\mathbb{N}$? It is easy to see that metrizability, separability, and second countability are being preserved under this construction. But the remaining 3 of the above 6 properties may still fail to hold:

Disaster 4.61. If $\mathbf{CC}(\mathbb{Z})$ fails there exists a compact Hausdorff space Z with $\dim Z = 0$ such that $Z \times \mathbb{N}$, the sum of countably many copies of Z, is neither normal, nor Lindelöf nor has dimension 0.

Proof. Consider the space Y, constructed in the proof of Disaster 4.60. Being locally compact, Y has a 1–point Hausdorff compactification $Z = Y \cup \{\infty\}$. Obviously $\dim Z = 0$. Consider the first projection $\pi_Z \colon Z \times \mathbb{N} \to Z$. If $Z \times \mathbb{N}$ would be normal, then the disjoint closed sets $A^* = \{(a_n, n) \mid n \in \mathbb{N}\}$ and $B^* = \{(b_n, n) \mid n \in \mathbb{N}\}$ would have disjoint neighborhoods U^* and V^* in $Z \times \mathbb{N}$. Consequently the sets $U = \bigcup_n \pi_Z[(Y_n \times \{n\}) \cap U^*]$ and $V = \bigcup_n \pi_Z[(Y_n \times \{n\}) \cap V^*]$ would be disjoint neighborhoods of $A = \{a_n \mid n \in \mathbb{N}\}$ and $B = \{b_n \mid n \in \mathbb{N}\}$ in Y, which — according to Claim 1 in the proof of Disaster 4.60 — cannot happen. Thus $Z \times \mathbb{N}$ fails to be normal, hence also to have dimension 0. Moreover, $Z \times \mathbb{N}$ fails to be Lindelöf since the open cover

$$\{(Z \backslash Y_n) \times \{n\} \mid n \in \mathbb{N}\} \cup \{(Y_n \backslash \{x\}) \times \{n\} \mid n \in \mathbb{N} \text{ and } x \in X_n\}$$

has no countable subcover.

In Theorem 3.8 we have shown that the sum of countably many copies of a one–point space, i.e., a countable discrete space, is Lindelöf if and only if $\mathbf{CC}(\mathbb{R})$ holds. Thus the Lindelöf property can get destroyed by this process. The following result describes the general situation:

Theorem 4.62. [96] *Equivalent are:*

1. *Countable sums of Lindelöf spaces are Lindelöf.*
2. *The sum of countably many copies of a compact Hausdorff space is Lindelöf.*
3. *The sum of a compact Hausdorff space with a countable discrete space is Lindelöf.*
4. **CC**.

Proof. **(1)** $\Rightarrow$ **(2)** is obvious.
(2) $\Rightarrow$ **(3)** Let X be a compact Hausdorff space and let N be a countable discrete space. Then the sum $X + N$ is homeomorphic to a closed subspace of

[96] [Bru82], [Her2002], [Ker200?].

the sum of countably many copies of the compact Hausdorff space $X + \{0\}$, the sum of X with a one–point space. Thus (3) follows from (2).

(3) ⇒ (4) Let (X_n) be a sequence of non–empty sets. Let $X = \bigcup_n X_n \cup \{\infty\}$ be the one–point compactification of the discrete space $\bigcup_n X_n$, and let $\mathbb{N}$ be the discrete space of natural numbers. By (3), the sum $X + \mathbb{N}$ is Lindelöf. Thus the open cover

$$\mathfrak{U} = \{X\} \cup \{\{n, x\} \mid n \in \mathbb{N} \text{ and } x \in X_n\}$$

contains a countable subcover $\{U_n \mid n \in \mathbb{N}\}$.

For each $n \in \mathbb{N}$ define

$$n^* = \min\{m \in \mathbb{N} \mid n \in U_m\}.$$

Then $U_{n^*} = \{n, x_n\}$ for a unique element x_n of X_n. Thus $(x_n) \in \prod_n X_n$.

(4) ⇒ (1) Let $X = \sum_n X_n$ be a countable sum of (pairwise disjoint) Lindelöf spaces, and let $\mathfrak{U}$ be an open cover of X. For each n, $\mathfrak{V}_n = \{U \cap X_n \mid U \in \mathfrak{U}\}$ is an open cover of X_n, thus it contains a countable subcover $\{V_m \mid m \in \mathbb{N}\}$ of X_n. For each $m \in \mathbb{N}$ the set $\mathcal{U}_m = \{U \in \mathfrak{U} \mid U \cap X_n = V_m\}$ is not empty. So, by **CC**, there exists a sequence $(U_m)_{m\in\mathbb{N}}$ in $\mathfrak{U}$ with $U_m \cap X_n = V_m$, hence with $X_n \subseteq \bigcup_m U_m$. Using **CC** again, we obtain a sequence $(\mathfrak{U}_n)$ of countable subsets of $\mathfrak{U}$ with $X_n \subseteq \bigcup \mathfrak{U}_n$. Using **CC** a third time, we conclude that $\mathfrak{V} = \bigcup_n \mathfrak{U}_n$ is a countable subset of $\mathfrak{U}$ that covers $X = \bigcup_n X_n$.

Remark 4.63. Observe, that by the above result, even the sum of two Lindelöf spaces may fail to be Lindelöf. In fact, if **CC**($\mathbb{R}$) holds, but **CC** fails[97], then there exists a compact Hausdorff space (even a one–point compactification of a discrete space) whose sum with the discrete Lindelöf space $\mathbb{N}$ of natural numbers fails to be Lindelöf.

Theorem 4.64. [98] *Equivalent are:*

1. *Countable sums of separable spaces are separable.*
2. **CC**.

Proof. **(1) ⇒ (2)** Let (X_n) be a sequence of non–empty sets. Assume w.l.o.g. that the X_n's are pairwise disjoint. Consider each X_n as an indiscrete, hence separable, topological space. By (1), there exists a sequence (d_n) in $\bigcup_{n\in\mathbb{N}} X_n$ such that $\{d_n \mid n \in \mathbb{N}\}$ is dense in $\sum_{n\in\mathbb{N}} X_n$. Define

$$x_n = d_{\min\{m\in\mathbb{N} \mid d_m \in X_n\}}.$$

[97] E.g., in Fraenkel's First Model A7 (N1 in [HoRu98]).

[98] [Ker200?], [KeTa2004].

Then $(x_n) \in \prod_{n\in\mathbb{N}} X_n$.

(2) $\Rightarrow$ **(1)** Let (X_n) be a sequence of (non-empty and pairwise disjoint) separable spaces. Then, for each $n \in \mathbb{N}$, the set D_n of all maps $f\colon \mathbb{N} \to X_n$ such that $f[\mathbb{N}]$ is dense in X_n, is non–empty. By (2), there exists some $(f_n) \in \prod_{n\in\mathbb{N}} D_n$. Thus $\bigcup_{n\in\mathbb{N}} f_n[\mathbb{N}]$ is a countable, dense subset of $\sum_{n\in\mathbb{N}} X_n$.

For a corresponding result about countable sums of normal spaces we need the following lemma whose interesting combinatorial proof is too long to be included here:

Lemma 4.65. [99] *Let $(X_i)_{i\in I}$ be a family of infinite sets. Let $\mathcal{P}^0_{\text{fin}}X_i$ be the set of all non–empty, finite subsets of X_i, and let M_i be the set consisting of all pairs $(\mathfrak{A}, \mathfrak{B})$ in $(\mathcal{P}^0_{\text{fin}}X_i)^2$ with $\mathfrak{A} \neq \emptyset \neq \mathfrak{B}$ and $A \cap B = \emptyset$ for each $A \in \mathfrak{A}$ and $B \in \mathfrak{B}$. Then there exists a family $(f_i)_{i\in I}$ of functions $f_i\colon M_i \to \mathcal{P}^0_{\text{fin}}X_i$.*

Theorem 4.66. [100] *Equivalent are*

1. *Countable sums of normal spaces are normal.*
2. **CMC**.

Proof. **(1)** $\Rightarrow$ **(2)** Let $(X_n)_{n\in\mathbb{N}}$ be a sequence of non–empty sets. To establish **CMC** we may assume w.l.o.g. that all the X_n's are infinite. For each n construct a normal topological space (Y_n, τ_n) as follows:

Y_n the disjoint union of $\mathcal{P}^0_{\text{fin}}X_n$ and a 2–element set $\{a_n, b_n\}$.

τ_n consists of all subsets A of Y_n which satisfy the following two properties:

1. If $a_n \in A$, then there exists some $F \in \mathcal{P}^0_{\text{fin}}X_n$ such that $F^{\uparrow} = \{G \in \mathcal{P}^0_{\text{fin}}X_n \mid F \subseteq G\} \subseteq A$.
2. If $b_n \in A$, then there exists some $F \in \mathcal{P}^0_{\text{fin}}X_n$ such that $F^{*} = \{G \in \mathcal{P}^0_{\text{fin}}X_n \mid G \cap F = \emptyset\} \subseteq A$.

Assume w.l.o.g. that the Y_n's are pairwise disjoint and form the sum $Y = \sum_{n\in\mathbb{N}} Y_n$. By (1), Y is normal. Thus the disjoint closed sets $A = \{a_n \mid n \in \mathbb{N}\}$ and $B = \{b_n \mid n \in \mathbb{N}\}$ have disjoint open neighborhoods U and V.

Define $\mathfrak{A}_n = \{F \in \mathcal{P}^0_{\text{fin}}X_n \mid F^{\uparrow} \subseteq U\}$ and
$\mathfrak{B}_n = \{F \in \mathcal{P}^0_{\text{fin}}X_n \mid F^{*} \subseteq V\}$.

Then the $\mathfrak{A}_n$'s and $\mathfrak{B}_n$'s are non–empty and satisfy $F \cap G \neq \emptyset$ for each $F \in \mathfrak{A}_n$ and $G \in \mathfrak{B}_n$. Thus, for each $n \in \mathbb{N}$, the pair $(\mathfrak{A}_n, \mathfrak{B}_n)$ belongs to M_n, as defined in Lemma 4.65. Let $(f_n)_{n\in\mathbb{N}}$ be a sequence of functions $f_n\colon M_n \to \mathcal{P}^0_{\text{fin}}X_n$. Then, for each n, $F_n = f_n(\mathfrak{A}_n, \mathfrak{B}_n)$ is a non–empty, finite subset of X_n. This establishes **CMC**.

(2) $\Rightarrow$ **(1)** Let $X = \sum_{n\in\mathbb{N}} X_n$ be a countable sum of (pairwise disjoint) normal spaces and let A and B be disjoint closed subsets of X. Then, for each

---

[99] [HKRR98]

[100] [HKRR98], [HKRR98a].

$n \in \mathbb{N}$, the sets $A_n = A \cap X_n$ and $B_n = B \cap X_n$ are disjoint closed subsets of X_n. Thus, by normality of the X_n's, for each $n \in \mathbb{N}$ the set P_n, consisting of all pairs (U, V) of disjoint open neighborhoods of A_n and B_n in X_n, is not empty. Denote the first and second projection of f_n by π_n^1 resp. π_n^2. By **CMC** there exists a sequence $(f_n)_{n \in \mathbb{N}}$ of non–empty, finite subsets F_n of $\mathcal{P}_n$. Thus, for each $n \in \mathbb{N}$, the sets $U_n = \bigcap \pi_n^1[f_n]$ and $V_n = \bigcap \pi_n^2[f_n]$ are disjoint open neighborhoods of A_n and B_n in X_n. Consequently $U = \bigcup_{n \in \mathbb{N}} U_n$ and $V = \bigcup_{n \in \mathbb{N}} V_n$ are disjoint open neighborhoods of A and B in X. Thus X is normal.

Exercises to Section 4.7:

E 1. Show that countable sums and countable products of metric spaces are metrizable.

E 2. Show the equivalence of the following conditions:
(1) Countable sums of metrizable spaces are metrizable.
(2) Countable products of metrizable spaces are metrizable.

E 3. Show that the product $\prod_n Y_n$ of the Y_n's, constructed in the proof of Disaster 4.60, is not metrizable.

E 4. Show that the spaces $\sum_n Y_n$ and $Z \times \mathbb{N}$, constructed in the proof of Disaster 4.60, respectively 4.61, are orderable[101].

E 5. [102] Show that, whenever **OP** holds and **KW** fails[103], there exists an orderable topological space that is a sum of normal spaces but fails to be normal itself.

E 6. Show the equivalence of:
(1) Finite sums of indiscrete spaces are Alexandroff–Urysohn–compact.
(2) Finite sums of Alexandroff–Urysohn–compact spaces are Alexandroff–Urysohn–compact.
(3) **AC**.

E 7. [104] Show the equivalence of the following conditions:
(1) Sums of normal spaces are normal.
(2) **AC**.
[Hint: Proceed as in the proof of Theorem 4.66 by using Lemma 4.65, and use Theorem 2.4.]

---

[101] A topological space (X, τ) is called *orderable* iff there exists a linear order (= chain) on X that induces the topology τ.

[102] [Kro86]

[103] E.g., in Howard–Rubin's First Model A3 (N38 in [HoRu98]).

[104] [HKRR98], [HKRR98a].

E 8. [105] Show the equivalence of the following conditions:
(1) Countable sums of Lindelöf metric spaces are separable.
(2) Countable products of Lindelöf metric spaces are separable.

E 9. [106] Show the equivalence of the following conditions:
(1) Countable sums of compact metric spaces are separable.
(2) Countable products of compact metric spaces are separable.
(3) Countable products of compact metric spaces are compact.
(4) Compact metric spaces are separable.
(5) Countable products of non–empty compact metric spaces are non–empty.

E 10. [107] Show the equivalence of the following conditions:
(1) Countable sums of Lindelöf metric spaces are Lindelöf.
(2) Countable sums of Lindelöf metric spaces are hereditarily Lindelöf.
(3) Countable products of Lindelöf metric spaces are hereditarily Lindelöf.

4.8 Disasters in Topology II: Products (The Tychonoff and the Čech–Stone Theorem)

The theorem just proved [the Tychonoff Theorem] *can lay good claim to being the most important theorem in general (nongeometric) topology.*

S. Willard[108]

The next theorem [the Tychonoff Theorem] *is fundamental in this context and is also one of the most important theorems of general topology.*

R. Engelking[109]

The Tychonoff Product Theorem *concerning the stability of compactness under formation of topological products may well be regarded as the single most important theorem of general topology.*

H. Herrlich and G.E. Strecker[110]

105 [KeTa2005]
106 [KeTa2005]
107 [KeTa2005]
108 [Wil70]
109 [Eng89]
110 [HeSt97a]

Alas, disaster strikes again:

Disaster 4.67. Products of compact spaces may fail to be compact.

In fact, nothing less than **AC** itself is needed to prove the Tychonoff Theorem:

Theorem 4.68. [111] *Equivalent are:*

1. *The Tychonoff Theorem: Products of compact spaces are compact.*
2. **AC**.

Proof. **(1)** $\Rightarrow$ **(2)** Let $(X_i)_{i\in I}$ be a family of non–empty sets and let ∞ be an element, not contained in $\bigcup\limits_{i\in I} X_i$. Define compact topological spaces (Y_i, τ_i) by $Y_i = X_i \cup \{\infty\}$ and $\tau_i = \{\emptyset, Y_i, \{\infty\}\}$. By (1), the space $P = \prod\limits_{i\in I}(Y_i, \tau_i)$ is compact. For each $i \in I$, the set $A_i = \pi_i^{-1}[X_i]$ is a non–empty, closed subset of P (where π_i denotes the i–th projection). The collection $\mathfrak{A} = \{A_i \mid i \in I\}$ has the finite intersection property. Thus $\bigcap\limits_{i\in I} A_i \neq \emptyset$ by compactness of P. Since $\bigcap\limits_{i\in I} A_i = \prod\limits_{i\in I} X_i$, **AC** follows.

(2) $\Rightarrow$ **(1)** See any book on general topology.

The situation gets only slightly more pleasant, if we restrict attention to Hausdorff spaces or even further to *Hilbert cubes*, i.e., products of the form $[0,1]^I$, or to *Cantor cubes*, i.e., products of the form $\mathbf{2}^I$, where $\mathbf{2}$ is the discrete space with underlying set $\{0,1\}$. Still, disasters cannot be avoided. Recall, however, that $[0,1]^{\mathbb{N}}$ and $\mathbf{2}^{\mathbb{N}}$ are compact (see Theorem 3.13).

Disaster 4.69.
1. Products of compact Hausdorff spaces may fail to be compact.
2. Hilbert cubes $[0,1]^I$ may fail to be compact.
3. Cantor cubes $\mathbf{2}^I$ may fail to be compact.

Theorem 4.70. [112] *Equivalent are:*

1. *Products of compact Hausdorff spaces are compact.*
2. *Products of finite discrete spaces are compact.*
3. *Products of finite spaces are compact.*
4. Hilbert cubes $[0,1]^I$ *are compact.*
5. Cantor cubes $\mathbf{2}^I$ *are compact.*
6. **PIT**.
7. **UFT**.

---

111 [Kel50]

112 [RuSc54], [LoRy55], [Myc64a], [Her96].

Proof. By Theorem 4.37, the conditions (1), (2), (6), and (7) are equivalent.

Moreover the implications (**1**) $\Rightarrow$ (**4**) $\Rightarrow$ (**5**) and (**3**) $\Rightarrow$ (**2**) are straightforward.

(**5**) $\Rightarrow$ (**2**) Let $(X_i)_{i\in I}$ be a family of finite discrete spaces. For each $i \in I$ let

$f_i \colon X_i \to \mathbf{2}^{C(X_i,\mathbf{2})}$ be the canonical map. Then each f_i and thus

$$\prod f_i \colon \prod_{i\in I} X_i \to \prod_{i\in I} \mathbf{2}^{C(X_i,\mathbf{2})} \cong \mathbf{2}^{\biguplus C(X_i,\mathbf{2})}$$

are closed embeddings. Thus compactness of $\prod\limits_{i\in I} X_i$ follows from that of $\mathbf{2}^{\biguplus C(X_i,\mathbf{2})}$.

(**2**) $\Rightarrow$ (**3**) Let $(X_i)_{i\in I}$ be a family of finite spaces. For each $i \in I$, let Y_i be the discrete space with the same underlying set as X_i. Then each X_i is a continuous image of Y_i, and thus $\prod\limits_{i\in I} X_i$ is a continuous image of $\prod\limits_{i\in I} Y_i$. Thus compactness of the latter implies compactness of the former.

What happens if we restrict the number of factors and consider only countable products?

Disaster 4.71. Countable products of compact spaces may fail to be compact.

Proposition 4.72. [113] *Each of the following conditions implies the subsequent ones:*

1. **DC**.
2. *Countable products of compact spaces are compact.*
3. **CC**.

Proof. (**1**) $\Rightarrow$ (**2**) Let $(X_n)_{n\in\mathbb{N}}$ be a sequence of compact spaces and let $\mathcal{F}$ be a filter on $X = \prod\limits_{n\in\mathbb{N}} X_n$. A cluster point $x = (x_n)_{n\in\mathbb{N}}$ of $\mathcal{F}$ can be constructed as in the proof of Theorem 3.13, using **DC** on (Y, ϱ), where:

1. Y is the set of all triples $(n, (x_0, x_1, \ldots, x_n), \mathcal{G})$, consisting of an element n of $\mathbb{N}$, an element $(x_0, x_1, \ldots, x_n)$ of $\prod\limits_{i=0}^{n} X_i$ and a filter $\mathcal{G}$ on X such that $\mathcal{F} \subseteq \mathcal{G}$ and $\pi_i^{-1}[U] \in \mathcal{G}$ for each $i \in \{0, 1, \ldots, n\}$ and each neighborhood U of x_i in X_i (where π_i denotes the i–th projection).
2. ϱ is defined by:
$(n, (x_0, \ldots, x_n), \mathcal{G})\ \varrho\ (m, (y_0, \ldots, y_m), \mathcal{H})$ iff $m = n + 1, (x_0, \ldots, x_n) = (y_0, \ldots, y_n)$, y_m is a cluster point of the filter $\{G \subseteq X_m \mid \pi_m^{-1}[G] \in \mathcal{G}\}$, and $\mathcal{H}$ is the filter, generated by the set $\mathcal{G} \cup \{\pi_m^{-1}[U] \mid U$ is a neighborhood of y_m in $X_m\}$.

[113] [GoTr95]

(2) $\Rightarrow$ **(3)** Analogous to the proof of the corresponding implication in Theorem 4.68.

Remark 4.73. The implication (1) $\Rightarrow$ (2) of Proposition 4.72 is a proper one, since there exists a **ZF**–model[114] in which **PIT** and **CC**, and thus condition (2) hold (see Exercise E 4), but **DC** fails. It is not known whether the implication (2) $\Rightarrow$ (3) is also proper.

What happens, if we restrict things further by considering only countable products of finite (discrete) spaces?

Disaster 4.74. Countable products of finite spaces may fail to be compact.

Theorem 4.75. [115] *Equivalent are:*

1. *Countable products of finite spaces are compact.*
2. *Countable products of finite discrete spaces are compact.*
3. **CC**(fin).

Proof. **(1)** $\Leftrightarrow$ **(2)** Straightforward.

(2) $\Rightarrow$ **(3)** Analogous to the proof of the corresponding implication in Theorem 4.68.

(3) $\Rightarrow$ **(2)** Let $P = \prod_{n\in\mathbb{N}} X_n$ be the product of a sequence $(X_n)_{n\in\mathbb{N}}$ of finite discrete spaces, and let $\mathcal{F}$ be a filter on P. Then (3) implies, via Proposition 3.5, that $\bigcup_{n\in\mathbb{N}} X_n$ is at most countable, thus well–orderable. Now proceed as in the proof of Theorem 3.13, by choosing x_0 and each x_{n+1} as the smallest point with the desired properties.

Let us go even further and restrict attention, for a given natural number n, to countable products of spaces with n points each. Then for $n \in \{0, 1\}$ we are on safe and trivial ground, and even for $n = 2$ we remain safe: see Theorem 3.17. Moreover, Theorem 3.13 immediately implies, that for each $n \in \mathbb{N}$ the space $\mathbf{n}^{\mathbb{N}}$ is compact (where $\mathbf{n}$ is the discrete space with underlying set $\{0, 1, \ldots, n-1\}$[116]. However:

Disaster 4.76. Countable products of 3–element spaces may fail to be compact.

Theorem 4.77. [117] *For each $n \in \mathbb{N}$ the following conditions are equivalent:*

1. *Countable products of (discrete) spaces with at most $n+1$ points are compact.*

---

[114] Howard–Rubin's First Model A3 (N38 in [HoRu98]). See also [HoRu96].
[115] [Kro81]
[116] See also [Bru84].
[117] [HeKe2000]

2. $\mathbf{CC}(\leq n)$.
3. $\mathbf{CUT}(n)$.

Proof. **(2)** $\Leftrightarrow$ **(3)** See Exercises to Section 3.1, E 1.

(1) $\Rightarrow$ **(2)** Analogous to the proof of the corresponding implication in Theorem 4.68.

(2) $\Rightarrow$ **(1)** Let $P = \prod_{m\in\mathbb{N}} X_m$ be the product of a sequence of spaces X_m with at most $n+1$ points each.

Case 1: $P = \emptyset$. Then P is compact.

Case 2: $P \neq \emptyset$. Let $(x_m)_{m\in\mathbb{N}}$ be an element of P. Then, for each $m \in \mathbb{N}$, the set $Y_m = X_m \setminus \{x_m\}$ contains at most n elements. Thus, by (3), the set $Y = \bigcup_{m\in\mathbb{N}} Y_m$ is at most countable. Consequently, $\bigcup_{m\in\mathbb{N}} X_m = Y \cup \{x_m \mid m \in \mathbb{N}\}$ is at most countable, thus well–orderable. Now proceed as in the proof of Theorem 3.13 resp. Theorem 4.75.

Remark 4.78. Observe that $\mathbf{CC}(2)$ may fail in **ZF**[118].

Conclusion: In **ZF** the Tychonoff Theorem breaks down completely. However, there is still hope: As we have seen in Section 3.2, the compactness concept splits in **ZF** in various, no longer equivalent, variants. Do some of these variants behave any better?

Disaster 4.79. Products of ultrafilter–compact spaces may fail to be ultrafilter–compact.

In fact, the Tychonoff Theorem for ultrafilter–compact spaces holds if and only if either **AC** holds or **AC** fails badly. In between these extremes it fails:

Theorem 4.80. [119] *Equivalent are:*

1. *Products of ultrafilter–compact spaces are ultrafilter–compact.*
2. *Either* **AC** *holds or* **WUF***(?) fails, i.e., there are no free ultrafilters*[120].

Let us postpone the proof of Theorem 4.80 until after Remark 4.83 and first consider the Hausdorff case. Here the sky brightens:

4.81. Tychonoff Theorem for ultrafilter–compact Hausdorff spaces: Products of ultrafilter compact Hausdorff spaces are ultrafilter–compact.

Proof. Let $P = \prod_{i\in I} X_i$ be the product of a family of ultrafilter–compact Hausdorff spaces, and let $\mathcal{U}$ be an ultrafilter on P. Then, for each $i \in I$, the set $\{A \subseteq X_i \mid \pi_i^{-1}[A] \in \mathcal{U}\}$ is an ultrafilter on X_i, and thus converges to a unique point x_i in X_i. Consequently $\mathcal{U}$ converges to the point $(x_i)_{i\in I}$ in P.

---

118 E.g., in Cohen's Second Model (M7 in [HoRu98]) and in Fraenkel's Second Model (N2(2) in [HoRu98]).

119 [Her96]

120 **WUF**(?) fails, e.g., in the Feferman–Blass Model (M15 in [HoRu98]) and in Pincus–Solovay's Model A6 (M27 in [HoRu98]). See also [Blass77].

4.82. Čech–Stone Theorem for ultrafilter–compact Hausdorff spaces: [121] The ultrafilter–compact Hausdorff spaces form an epireflective subcategory of the category **Haus** of Hausdorff spaces and continuous maps.

Proof. Immediate from the fact that a full subcategory **A** of **Haus** is epireflective in **Haus** if (and only if) **A** is closed under the formation of products and closed subspaces. For details see, e.g., [Her68] or [AHS2004].

Remark 4.83. Even though the Čech–Stone Theorem holds for ultrafilter–compact Hausdorff spaces, the situation is not quite as satisfactory as in the **ZFC**–setting. In the latter there exist many detailed descriptions[122] of the Čech–Stone compactification βX (= the compact Hausdorff reflection) of X, at least for completely regular spaces, whereas in the **ZF** setting no detailed description of the ultrafilter–compact Hausdorff reflection is known — an exception being the discrete case. If X is a discrete space, then its *Wallman extension* ωX is an ultrafilter–compact Hausdorff reflection of X. Here follows a simple description of ωX: For each free ultrafilter $\mathcal{U}$ on X, add to X a point $p_{\mathcal{U}}$. Thus the underlying set of ωX has the form $X \cup \{p_{\mathcal{U}} \mid \mathcal{U}$ is a free ultrafilter on $X\}$.

For each subset A of X, define

$$A^* = A \cup \{p_{\mathcal{U}} \mid A \in \mathcal{U}\}.$$

Then $\{A^* \mid A \subseteq X\}$ is a base for the topology of ωX.

Now we are ready to present a
Proof of Theorem 4.80: **(1)** $\Rightarrow$ **(2)** Assume that there exists a free ultrafilter $\mathcal{U}$ on some set X. To show that **AC** holds, let $(X_i)_{i\in I}$ be a family of non–empty sets. Let ωX be the Wallman extension of the discrete space with underlying set X. Assume for simplicity that the underlying set of ωX is disjoint from each X_i. For each $i \in I$, consider space Y_i; obtained from ωX via replacement of the point $p_{\mathcal{U}}$ by the set X_i, considered as an indiscrete subspace of Y_i. Then each Y_i is ultrafilter–compact and the ultrafilter $\mathcal{U}_i$, generated by $\mathcal{U}$, has X_i as its set of limit points. By (1), the product space $Y = \prod_{i\in I} Y_i$ is ultrafilter compact. Let $\Delta\colon X \to Y$ be the diagonal embedding. Then in Y the filter $\mathcal{V}$, generated by $\Delta[\mathcal{U}]$, is an ultrafilter, and thus converges to some point y. Consequently, for each $i \in I$, the filter $\mathcal{U}_i = \{A \subseteq X_i \mid \pi_i^{-1}[A] \in \mathcal{V}\}$ converges in Y_i to $\pi_i(y)$. This implies $\pi_i(y) \in X_i$, thus $y \in \prod_{i\in I} X_i$. So **AC** holds.

(2) $\Rightarrow$ (1) If **AC** holds, then (1) follows as in the proof of Theorem 4.81. If **WUF**(?) fails, then every space is ultrafilter–compact, and thus (1) holds trivially.

Next, let us turn our attention to Tychonoff–compactness. As we will see, here the situation concerning the Tychonoff Theorem and the Čech–Stone

121 [Her96]

122 See, e.g., [HeSt97], where 25 different constructions of βX are exhibited.

Theorem (and also the theory of rings of continuous functions) is as pleasant as in the **ZFC**–setting[123].

4.84. Tychonoff Theorem for Tychonoff–compact spaces:[124] Products of Tychonoff–compact spaces are Tychonoff–compact.

Proof. Let $(X_i)_{i\in I}$ be a family of Tychonoff–compact spaces. Then, for each $i \in I$, the canonical map

$$j_i \colon X_i \to [0,1]^{C(X_i,[0,1])},$$

defined by $\pi_f \circ j_i = f$ for each $f \in C(X_i,[0,1])$, is a closed embedding (see Exercises to Section 3.3, E 4). Consequently, the product map

$$\prod_{i\in I} j_i \colon \prod_{i\in I} X_i \longrightarrow \prod_{i\in I} [0,1]^{C(X_i,[0,1])} \cong [0,1]^{\biguplus C(X_i,[0,1])}$$

is a closed embedding, too, and thus $\prod\limits_{i\in I} X_i$ is Tychonoff–compact.

Observe that the above theorem could have been included into Section 3.2, since among all the closed embeddings of a Tychnonoff–compact space into powers $[0,1]^I$ of $[0,1]$ there exists (according to Exercises to Section 3.3, E 4) a distinguished one.

4.85. Čech–Stone Theorem for Tychonoff–compact spaces[125] Tychonoff–compact spaces form an epireflective subcategory of the category **Tych** of completely regular spaces and continuous maps. In particular, every completely regular space X can be densely embedded into a Tychonoff–compact space βX (its *Čech–Stone compactification*) such that the following equivalent properties are satisfied:

1. Every continuous map $X \to [0,1]$ can be extended to a continuous map $\beta X \to [0,1]$.
2. Every bounded, continuous map $X \to \mathbb{R}$ can be extended to a continuous map $\beta X \to \mathbb{R}$.
3. Every continuous map $X \to C$ from X into some Tychonoff–compact space C can be extended to a continuous map $\beta X \to C$.

Proof. Let X be a completely regular space. Then the canonical map

$$j \colon X \to [0,1]^{C(X,[0,1])}$$

is an embedding. Factor j through the closure $\beta X = \mathrm{cl}[j[X]]$ of its image $j[X]$ in $[0,1]$

$$X \xrightarrow{j} [0,1]^{C(X,[0,1])} \quad = \quad X \xrightarrow{\beta} \beta X \hookrightarrow [0,1]^{C(X,[0,1])}.$$

Then the dense embedding $\beta \colon X \to \beta X$ has the desired properties[126].

---

[123] See [Com68], [Sal74], [Her96], [BeHe99].

[124] [Com68], [Sal74].

[125] [Com68], [Sal74]

[126] See, e.g., [Her68] for details.

Remark 4.86. Alternative constructions of the Tychonoff–compact reflection $X \to \beta X$ can be found in [Com68] (via rings of continuous functions), in [Cha72] and in [Sal74] (via zero ultrafilters and the Wallman construction).

Theorem 4.87. [127] *The Tychonoff–compact spaces form the* epireflective hull[128] *of the completely regular compact spaces in the category* **Tych**.

Proof. Let **A**, resp. **B**, be the class of all completely regular compact, resp. all Tychonoff–compact, spaces and let **C** be the epireflective hull of **A** in **Tych**. For each X in **A** the canonical map

$$j\colon X \to [0,1]^{C(X,[0,1])}$$

is a closed embedding. (Complete regularity implies that j is an embedding, compactness implies that $j[X]$ is closed in $[0,1]^{C(X,[0,1])}$, see Exercises to Section 3.3, E 2). Thus $\mathbf{A} \subseteq \mathbf{B}$ and thus $\mathbf{C} \subseteq \mathbf{B}$. Since $[0,1]$ belongs to **A**, all closed subspaces of powers of $[0,1]$, i.e., all elements of **B**, belong to **C**. Thus $\mathbf{B} \subseteq \mathbf{C}$. Consequently $\mathbf{B} = \mathbf{C}$.

Finally, let us turn our attention to Alexandroff–Urysohn–compact spaces. Here the Tychonoff Theorem breaks down completely.

Disaster 4.88. Even finite products of Alexandroff–Urysohn–compact spaces may fail to be Alexandroff–Urysohn–compact.

Theorem 4.89. [129] *Equivalent are:*

1. *Products of Alexandroff–Urysohn–compact spaces are Alexandroff–Urysohn–compact.*
2. *Finite products of Alexandroff–Urysohn–compact spaces are Alexandroff–Urysohn–compact.*
3. *Cantor cubes $\mathbf{2}^I$ are Alexandroff–Urysohn compact.*
4. **AC**.

Proof. Obviously (**1**) implies (**2**) and (**3**).

(**2**) $\Rightarrow$ (**4**) By Theorem 4.20 it suffices to show that any two cardinals are comparable w.r.t. $\leq$. Let A and B be infinite sets with cardinals $|A| = a$ and $|B| = b$. Then the indiscrete space X with underlying set $A \cup B$, and the discrete space **2** (with underlying set $\{0,1\}$) are both Alexandroff–Urysohn–compact, and thus — by (2) — so is their product $X \times \mathbf{2}$. Thus the set $C = (A \times \{0\}) \cup (B \times \{1\})$ has a complete accumulation point (x, y) in $X \times \mathbf{2}$. If $y = 0$, then $(A \cup B) \times \{0\}$ is a neighborhood of (x, y) that meets C in

127 [Her96]

128 The *epireflective hull* of **A** in **B** is the smallest epireflective subcategory of **B** that contains **A** (provided that such an entity exists).

129 [Her68]

$A \times \{0\}$. Thus $a = |A| = |A \times \{0\}| = |C| = |A \cup B| = a + b$, and hence $b \leq a$. Likewise $y = 1$ implies $a \leq b$. Consequently a and b are comparable.

(3) $\Rightarrow$ **(4)** As above, consider arbitrary infinite sets A and B and let ∞ be not contained in $A \cup B$. Form $I = \{\infty\} \cup A \cup B$ and consider the Cantor cube $\mathbf{2}^I$. For a subset C of I, let $\chi_C \colon I \to \mathbf{2}$ be the associated characteristic function, defined by $\chi_C(i) = \begin{cases} 1, \text{ if } i \in C \\ 0, \text{ if } i \notin C \end{cases}$.

By (3) the

$$\{\chi_{\{a\}} \mid a \in A\} \cup \{\chi_{\{\infty,b\}} \mid b \in B\}$$

has a complete accumulation point in $\mathbf{2}^I$. As above, this implies that the cardinal numbers of A and B are comparable with respect to $\leq$.

(4) $\Rightarrow$ **(1)** is well–known.

Exercises to Section 4.8:

E 1. Show that finite products of compact spaces are compact.
[Hint: Proceed as in the proof of Theorem 3.13].

E 2. Show that finite products of ultrafilter–compact spaces are ultrafilter–compact.

E 3. Show that **CC** implies that countable products of ultrafilter–compact spaces are ultrafilter–compact.

E 4. Show that **CC** and **PIT** together imply that countable products of compact spaces are compact.

E 5. [130] Let X be a dense subspace of a Tychonoff–compact space Y. Show that the following conditions are equivalent:

(1) Y is (up to homeomorphism) the Tychonoff–compact reflection βX of X.
(2) Any two disjoint zero sets in X have disjoint closures in Y.
(3) For any two zero sets A and B in X, we have $\mathrm{cl}_Y(A \cap B) = \mathrm{cl}_Y A \cap \mathrm{cl}_Y B$.
(4) Every point of Y is the limit of a unique zero–ultrafilter in X.

E 6. [131] Show the equivalence of the following conditions:

(1) Countable products of compact pseudometric spaces are compact.
(2) $X^{\mathbb{N}}$ is compact for each compact pseudometric space X.
(3) **CC**.

E 7. Show the equivalence of the following conditions:

(1) Products of 2–element spaces are compact
(2) Products of non–empty compact Hausdorff spaces are non–empty and compact.
(3) **PIT**.

[130] [Sal74]
[131] [HeKe2000a]

E 8. Show the equivalence of the following conditions:
(1) Hilbert cubes $[0,1]^I$ are Alexandroff–Urysohn–compact.
(2) Tychonoff–compact spaces are Alexandroff–Urysohn–compact.
(3) **AC**.

E 9. Show that **PIT** implies **AC**(fin).
[Hint: Use Theorem 4.70 and the proof of Theorem 4.68.]

E 10. [132] A topological space X is called *supercompact* iff there exists some $x \in X$ whose only neighborhood is X itself. Prove that:
(1) Every product of supercompact T_0–spaces is supercompact.
(2) **AC** holds iff every product of supercompact spaces is supercompact.

E 11. Show the equivalence of the following conditions:
(1) Products of compact T_1–spaces are compact.
(2) **AC**.

[Hint: For (1) $\Rightarrow$ (2) proceed as in the proof of Theorem 4.68 but enrich the topologies of the spaces (Y_i, τ_i) by substituting $\sigma_i = \tau_i \cup \{Y_i \setminus F \mid F$ finite $\}$ for τ_i.]

E 12. [133] Show the equivalence of the following conditions:
(1) In the product topology the closure $\mathrm{cl}(\prod_{i\in I} A_i)$ of a product of subsets A_i of X_i is equal to the product $\prod_{i\in I} \mathrm{cl}_i A_i$ of the closures of the A_i in X_i.
(2) **AC**.

[Hint: For (1) $\Rightarrow$ (2) proceed as in the proof of Theorem 4.68, but use indiscrete topologies $\tau_i = \{\emptyset, Y_i\}$.]

E 13. [134] Show the equivalence of the following conditions:
(1) $[0,1]^{\mathbb{R}}$ is compact.
(2) $\mathbf{2}^{\mathbb{R}}$ is compact.
(3) Products of finite subspaces of $\mathbb{R}$ are compact.
(4) **UFT**($\mathbb{N}$).

[132] [Ban93]
[133] [Sch92]
[134] [Ker2005]

4.9 Disasters in Topology III: Function Spaces (The Ascoli Theorem)

Mathematics is the art of deduction rather than a list of facts.
V.W. Marek and J. Mycielski[135]

The **Tychonoff Theorem** provides the foundation for several important results in topology and in functional analysis. One of these is the **Čech–Stone Theorem**, which has been treated in the previous section. Another one is the **Ascoli Theorem**, which will be analyzed in this section. It concerns the characterization of compactness among certain spaces of continuous functions.

Unfortunately — even in **ZFC** — the category **Top** of topological spaces and continuous maps as well as all its topological subcategories that are closed under the formation of squares and contain the Sierpiński–space[136], fail to be *cartesian closed*[137], i.e., none of these categories has *function spaces* which have all the categorically desirable properties. However, the set $C(X,Y)$ of all continuous functions from X to Y carries several canonical topologies. The *weak topology*, i.e., the one induced by the product topology on the space Y^X, is for many applications too coarse. A finer and more useful topology, coming close to satisfying the categorically desirable properties, is the *compact–open topology* τ_{co} which has as canonical subbase the set of all sets of the form

$$[K,U] = \{f \in C(X,Y) \mid f[K] \subseteq U\},$$

where K is compact in X and U is open in Y. The space $C_{co}(X,Y) = (C(X,Y),\ \tau_{co})$ coincides, for discrete spaces X, with the product space Y^X.

The Ascoli Theorem takes on various forms. For our purpose the following version seems appropriate.

Definition 4.90. *The* **Ascoli Theorem** *states that for every locally compact Hausdorff space X, for every metric space Y, and for every subspace F of $C_{co}(X,Y)$, the following conditions are equivalent:*

(a) F is compact.

(b) (α) For each $x \in X$ the set $F(x) = \{f(x) \mid f \in F\}$ is compact in Y.

(β) F is closed in the product space Y^X.

(γ) F is equicontinuous, *i.e.*

$$\forall x \in X \ \ \forall \epsilon > 0 \ \ \exists U \in \mathcal{U}(x) \ \ \forall f \in F \ \ \forall y \in U \ \ d(f(x), f(y)) < \epsilon,$$

[135] [MaMy2001]

[136] The Sierpiński space is the space with underlying set $2 = \{0,1\}$ and open sets $\emptyset$, $\{0\}$, and 2.

[137] See, e.g., [Her83].

where $\mathcal{U}(x)$ is the neighborhood–filter of x in X and d is the distance–function of Y.

Theorem 4.91. [138] *Equivalent are:*

1. *The Ascoli Theorem.*
2. **PIT**.

Proof. **(1)** $\Rightarrow$ **(2)** Let X be a set and let $\mathbf{2}$ be the discrete space with underlying set $\{0,1\}$. Consider X as a discrete space and $\mathbf{2}$ as a metric space with distance–function d determined by $d(0,1)=1$. Let F be $C(X,\mathbf{2})=\mathbf{2}^X$. Then condition (b) of the Ascoli Theorem is satisfied. Thus (1) implies that $\mathbf{2}^X$ is compact. Hence, by Theorem 4.70, **PIT** holds.

(2) $\Rightarrow$ **(1)** Let X, Y, and F be as specified in the Ascoli Theorem.

(a) $\rightarrow$ **(b,** α**)** Since F is compact in $C_{co}(X,Y)$, it is compact in Y^X. Thus, for each $x \in X$, its image $\pi_x[F] = F(x)$ under the projection map $\pi_x \colon Y^X \to Y$ is compact.

(a) $\rightarrow$ **(b,** β**)** Again, F is compact in the Hausdorff space Y^X, and thus (by Exercises to Section 3.3, E 2) closed in Y^X.

(a) $\rightarrow$ **(b,** γ**)** Choose $x \in X$ and $\epsilon > 0$. Then, for each $f \in F$, the set $B_f = \{y \in Y \mid d(f(x),y) < \frac{\epsilon}{2}\}$ is open in Y. Thus, by continuity of f and local compactness of X, there exists a compact neighborhood K_f of x with $f[K_f] \subseteq B_f$. Thus

$$U_f = F \cap [K_f, B_f] = \{g \in F \mid g[K_f] \subseteq B_f\}$$

is an open neighborhood of f in F. Consider the evaluation map $\omega\colon X \times F \to Y$, defined by $\omega(y,g) = g(y)$. Then the above implies $\omega[K_f \times U_f] \subseteq B_f$. Consider the collection $\mathcal{C}$ of all triples (f,K,U) with $f \in F$, K a neighborhood of x in X, and U an open neighborhood of f in F with $\omega[K \times U] \subseteq B_f$. Then, by the above

$$\mathfrak{U} = \{U \subseteq F \mid \exists f \in F \ \ \exists K \subseteq X \quad (f,K,U) \in \mathcal{C}\}$$

is an open cover of F. Thus, by (a), there exist finitely many members $U_1, \ldots, U_n$ of $\mathfrak{U}$ which cover F. For each $i \in \{1,\ldots,n\}$, select $f_i \in F$ and $K_i \subseteq X$ with $(f_i, K_i, U_i) \in \mathcal{C}$. Then $U = \bigcap_{i=1}^{n} K_i$ is a neighborhood of x in X.

Claim: $\forall f \in F \quad \forall y \in U \quad d(f(x), f(y)) < \epsilon.$

Proof. For $f \in F$ there exists some $i \in \{1,\ldots,n\}$ with $f \in U_i$. Thus $y \in U$ implies:

$$f(y) = \omega(y,f) \subseteq \omega[U \times U_i] \subseteq \omega[K_i \times U_i] \subseteq B_{f_i},$$

i.e., $d(f_i(x), f(y)) < \frac{\epsilon}{2}$. In particular, $x \in U$ implies $d(f_i(x), f(x)) < \frac{\epsilon}{2}$. Thus

[138] [Her97]

$$d(f(x), f(y)) \leq d(f(x), f_i(x)) + d(f_i(x), f(y)) < \epsilon.$$

Consequently F is equicontinuous.

(b) → (a) By (b, α), each $F(x)$ is a compact Hausdorff space. Thus, by condition (2) and Theorem 4.70, $\prod_{x \in X} F(x)$ is compact. By (b, β), F is closed in Y^X and thus in $\prod_{x \in X} F(x)$. Consequently F is compact with respect to the weak topology τ, i.e., when considered as a subspace of Y^X. So it remains to be shown that (b, γ) implies that τ equals the (generally finer) compact–open topology σ on F. For this purpose consider an element $V = [K, U] \cap F$ of the canonical subbase of σ, and let f be an element of V. It remains to be shown that V is a neighborhood of f in the weak topology. Since $f[K] \subseteq U$ and U is open in Y, we obtain for each $x \in K$

$$r_x = \inf\{d(f(x), y) \mid y \in (Y \setminus U)\} > 0.$$

Then $U_x = \{z \in X \mid d(f(x), f(z)) < \frac{r_x}{2}\}$ is an open neighborhood of x in X, and $\mathfrak{U} = \{U_x \mid x \in K\}$ is an open cover of K. By compactness of K there exist finitely many members $x_1, \ldots, x_n$ of K such that $K \subseteq \bigcup_{i=1}^{n} U_{x_i}$. Thus $r = \min\{r_{x_1}, \ldots, r_{x_n}\} > 0$, and for each $x \in K$ and each $y \in (Y \setminus U)$ the inequality $d(f(x), y) \geq \frac{r}{2}$ follows; in other words:

$x \in K$ and $d(f(x), y) < \frac{r}{2}$ imply $y \in U$. By equicontinuity of F there exists, for each $x \in X$, some neighborhood W of x in X such that:

$$(*) \qquad \forall g \in F \quad \forall z \in W \quad d(g(x), g(z)) < \tfrac{r}{4}.$$

Consider the set $\mathcal{C}$ of all pairs (x, W) with $x \in X$ and W an open neighborhood of x in X such that (*) holds. Then

$$\mathfrak{U} = \{W \subseteq X \mid \exists x \in K \ (x, W) \in \mathcal{C}\}$$

is an open cover of K. By compactness of K there exist finitely many members $W_1, \ldots, W_m$ in $\mathfrak{U}$ which cover K. For each $i \in \{1, \ldots, m\}$ select some x_i with $(x_i, W_i) \in \mathcal{C}$. Then

$$B = \{g \in F \mid d(f(x_i), g(x_i)) < \frac{r}{4} \text{ for } i = 1, \ldots, n\}$$

is a neighborhood of f in the weak topology τ.

Claim: $B \subseteq V$.

Proof. Consider $g \in B$. For each $x \in K$ there exists some $i \in \{1, \ldots, m\}$ with $x \in W_i$. This implies $d(g(x_i), g(x)) < \frac{r}{4}$ by (*). Since $g \in B$, the inequality $d(f(x_i), g(x_i)) < \frac{r}{4}$ holds. Thus:

$$d(f(x_i), g(x)) \leq d(f(x_i), g(x_i)) + d(g(x_i), g(x)) < \frac{r}{2}.$$

Consequently, $g(x) \in U$; hence $g[K] \subseteq U$; hence $g \in V$.
This completes the proof.

Observe that the implication (a) $\Rightarrow$ (b) of the Ascoli Theorem holds in **ZF**, and that for the reverse implication (b) $\Rightarrow$ (a) the only non–**ZF** requirement is that products of compact Hausdorff spaces are compact. Since products of ultrafilter–compact Hausdorff space are always ultrafilter–compact, we may conjecture that the Ascoli Theorem holds, if we replace the notion of *compactness*, wherever it occurs (hence particularly in the definition of the compact–open topology), by that of *ultrafilter–compactness.* Alas:

Theorem 4.92. [139] *Equivalent are:*

1. *The Ascoli Theorem w.r.t. ultrafilter–compactness.*
2. **PIT**.

Proof. **(1)** $\Rightarrow$ **(2)** By Theorem 4.70 it suffices to show that condition (1) implies that all Cantor cubes $\mathbf{2}^I$ are compact. Assume that, for some set I, the space $P = \mathbf{2}^I$ fails to be compact. Consider I as a discrete space and $\mathbf{2}$ as a metric space with $d(0,1) = 1$. Then there exists a filter $\mathcal{F}$ on P without a cluster point. Thus the set X of all clopen[140] members of $\mathcal{F}$ has the finite–intersection–property, but empty intersection. For each $A \in X$, the map $f_A \colon P \to \mathbf{2}$, defined by $f_A(x) = \begin{cases} 1, \text{ if } x \in A \\ 0, \text{ if } x \notin A \end{cases}$, is continuous. Thus the family $(f_A)_{A\in X}$ induces a continuous map $f \colon P \to \mathbf{2}^X$. Let $F = f[P]$ be the image of P under f. A simple computation shows that f is an embedding. So F is homeomorphic to P and thus, by Theorem 4.81, F is ultrafilter–compact. Apply the Ascoli Theorem w.r.t. ultrafilter–compactness to X, considered as a discrete space, $Y = \mathbf{2}$, and F. Then condition (a) is satisfied. However condition (b β) fails, since the point $p = (1)_{A\in X}$ of $\mathbf{2}^X$ whose coordinates are all 1, belongs to the closure of F in $\mathbf{2}^X$, since X has the finite–intersection–property, but not to F, since X has empty intersection. Thus (1) fails, a contradiction.

(2) $\Rightarrow$ **(1)** By Theorems 3.22 and 4.37, **PIT** implies that compact = ultrafilter–compact. Thus (1) follows from Theorem 4.91.

Since for Tychonoff–compact spaces the Tychonoff Theorem and the Čech–Stone Theorem hold, we may expect the Ascoli Theorem for Tychonoff–compact spaces to hold as well. Alas:

---

[139] [Her97a]
[140] *Clopen* means *closed and open.*

Theorem 4.93. [141] *Equivalent are:*

1. *The Ascoli Theorem w.r.t. Tychonoff-compactness.*
2. **PIT**.

Proof. Since the proof of Theorem 4.93 parallels that of Theorem 4.92 it is left as an exercise (see Exercise E 1).

In view of the above facts the following result is no longer surprising:

Theorem 4.94. [142] *Equivalent are:*

1. *The Ascoli Theorem w.r.t. Alexandroff–Urysohn–compactness.*
2. **AC**.

Proof. **(1)** ⇒ **(2)** By Theorem 4.89 it suffices to show that all Cantor–cubes $\mathbf{2}^I$ are Alexandroff–Urysohn–compact. Let X be the discrete space with underlying set I, Y be the space $\mathbf{2}$, considered as a metric space with $d(0,1) = 1$, and $F = \mathbf{2}^X$. Then condition (b) of the Ascoli–Theorem w.r.t. Alexandroff–Urysohn–compactness is trivially satisfied. Thus, by condition (1), condition (a) holds as well, i.e., $\mathbf{2}^I$ is Alexandroff–Urysohn–compact.

(2) ⇒ **(1)** Under **AC**, Alexandroff–Urysohn–compactness agrees with compactness. Thus (1) follows from Theorem 4.91.

Sadly enough our resumé of the above results is this:

Disaster 4.95. The Ascoli Theorem may fail under each of the interpretations of *compactness* given above.

So there seems no hope left to salvage the Ascoli Theorem. Recall however that the classical form of the Ascoli Theorem is more restricted than the one formulated in 4.90. Can the former perhaps be saved?

Definition 4.96. *The* **Classical Ascoli Theorem** *states that for any set F of continuous maps $f\colon \mathbb{R} \to \mathbb{R}$ the following conditions are equivalent:*

(a) Each sequence (f_n) in F has a subsequence $f_{(\nu(n))}$ that converges continuously[143] *to some map g (not necessarily in F), i.e.,*

---

[141] [Her97a]

[142] [Her97a]

[143] Observe:

a) If (f_n) converges continuously to g, then g is continuous.

b) If (f_n) converges locally uniformly to g, then (f_n) converges continuously to g.

c) If (f_n) converges continuously to g, then (f_n) converges pointwise to g.

$$\forall x \in \mathbb{R} \quad \forall (x_n) \in \mathbb{R}^{\mathbb{N}} \quad ((x_n) \to x \Rightarrow (f_{\nu(n)}(x_n)) \to g(x)).$$

(b) (α) For each $x \in \mathbb{R}$ the set $F(x) = \{f(x) \mid f \in F\}$ is bounded.
(β) F is equicontinuous.

Disaster strikes even here:

Theorem 4.97. [144] *Equivalent are:*

1. *The Classical Ascoli Theorem.*
2. $\mathbf{CC}(\mathbb{R})$.

Proof. **(1)** $\Rightarrow$ **(2)** By Theorem 3.8 it suffices to show that each unbounded subset B of $\mathbb{R}$ contains an unbounded sequence. For this purpose, consider for each $b \in B$ the constant map $f_b \colon \mathbb{R} \to \mathbb{R}$ with value b. Then the set $F = \{f_b \mid b \in B\}$ violates condition (b, α) of the Classical Ascoli Theorem. So, by (1) it violates condition (a) as well. Thus there exists a sequence (f_{b_n}) in F that has no continuously convergent subsequence. This fact implies that the sequence (b_n) is unbounded.

(2) $\Rightarrow$ **(1)** Observe first that every continuous map $f \colon \mathbb{R} \to \mathbb{R}$ is determined by its restriction $f|_{\mathbb{Q}} \colon \mathbb{Q} \to \mathbb{R}$ to the rationals. This implies that there are only $|\mathbb{R}^{\mathbb{Q}}| = (2^{\aleph_0})^{\aleph_0} = 2^{\aleph_0 \cdot \aleph_0} = 2^{\aleph_0} = |\mathbb{R}|$ continuous functions $f \colon \mathbb{R} \to \mathbb{R}$. Hence $\mathbf{CC}(\mathbb{R})$ implies that for every sequence (F_n) of non–empty sets of continuous functions $f \colon \mathbb{R} \to \mathbb{R}$ there exists some choice–sequence $(f_n) \in \prod_{n \in \mathbb{N}} F_n$.

With this fact in mind let us turn to the proof of the Classical Ascoli Theorem: Let F be a set of continuous maps $f \colon \mathbb{R} \to \mathbb{R}$.

(a) $\Rightarrow$ **(b,** α**)** Assume that (b, α) fails. Then there exists some $x \in \mathbb{R}$ such that $F(x)$ is unbounded. Thus, for each $n \in \mathbb{N}$, the set $F_n = \{f \in F \mid |f(x)| \geq n\}$ is not empty. Consequently there exists $(f_n) \in \prod_{n \in \mathbb{N}} F_n$. Obviously (f_n) has no continuously convergent subsequence. This contradicts condition (a).

(a) $\Rightarrow$ **(b,** β**)** Assume that (b, β) fails. Then there exists some $x \in \mathbb{R}$ and some $\epsilon > 0$ such that, for each $n \in \mathbb{N}$, the set

$$F_n = \{f \in F \mid \exists y \in \mathbb{R} \ \ |x - y| < \frac{1}{n+1} \quad \text{and} \quad |f(x) - f(y)| \geq \epsilon\}$$

is non–empty. Consequently there exists $(f_n) \in \prod_{n \in \mathbb{N}} F_n$.

Obviously (f_n) has no continuously convergent subsequence. This contradicts condition (a).

(b) $\Rightarrow$ **(a)** Let (f_n) be a sequence in F. Express the rationals as a sequence (r_n) and define, by induction, a sequence of pairs (a_n, s_n) with $a_n \in \mathbb{R}$ and $s_n = (g_m^n)_{m \in \mathbb{N}}$ a sequence in F as follows:

---

[144] [Rhi2001]

1. Let a_0 be the smallest cluster point of the sequence $(f_n(r_0))$.
Define $s_0 = (g_n^0)_{n \in N}$ by induction as a subsequence $(f_{\nu(n)})$ of (f_n) as follows:

$$\begin{aligned} &\text{a) } \nu(0) &&= \min\{m \in \mathbb{N} \mid f_m(r_0) - a_0| < 1\} \\ &\text{b) } \nu(n+1) &&= \min\{m \in \mathbb{N} \mid \nu(n) < m \text{ and } |f_m(r_0) - a_0| < \tfrac{1}{n+1}\}. \end{aligned}$$

Then $s_0 = (g_n^0) = (f_{\nu(n)})$ is a subsequence of (f_n), and $(g_n^0(r_0)) \to a_0$.
2. Let a_n and $s_n = (g_m^n)_{m \in \mathbb{N}}$ be defined. Let a_{n+1} be the smallest cluster point of the sequence $(g_m^n(r_{n+1}))_{m \in \mathbb{N}}$.
Define, as above via induction, $(s_{n+1}) = (g_m^{n+1})_{m \in \mathbb{N}}$ as a subsequence of $(s_n) = (g_m^n)_{m \in \mathbb{N}}$ such that $(g_m^{n+1}(r_{n+1}))_{m \in \mathbb{N}} \to a_{n+1}$.
Next, consider the diagonal sequence $s = (g_n^n)_{n \in \mathbb{N}}$. Then s is a subsequence of (f_n) and is cofinal with each of the sequences s_n. Thus, for each $n \in \mathbb{N}$, the sequence $(s(r_n)) = \left(g_m^m(r_n)\right)_{m \in \mathbb{N}}$ converges to a_n. Hence, for each $x \in \mathbb{Q}$, the sequence $s(x) = (g_m^m(x))_{m \in \mathbb{N}}$ converges in $\mathbb{R}$. Since $\mathbb{Q}$ is dense in $\mathbb{R}$, $G = \{g_m^m \mid m \in \mathbb{N}\}$ is equicontinuous, and $\mathbb{R}$ is complete, the familiar arguments imply that s converges locally uniformly and thus continuously to some map $g\colon \mathbb{R} \to \mathbb{R}$ with $g(r_n) = a_n$ for each $n \in \mathbb{N}$. Consequently (a) holds.

Thus disaster struck again. However, a simple analysis of the above proof immediately yields the following salvaged form of the Classical Ascoli Theorem:

4.98. Modified Ascoli Theorem[145] For sets F of continuous maps $f\colon \mathbb{R} \to \mathbb{R}$ the following conditions are equivalent:

(a) Each sequence in F has a subsequence that converges continuously to some map g (not necessarily in F).
(b) (α) For each $x \in \mathbb{R}$ and each countable subset G of F the set $G(x) = \{g(x) \mid g \in G\}$ is bounded.
(β) Each countable subset of F is equicontinuous.

Exercises to Section 4.9:

E 1. Prove Theorem 4.93.

E 2. Prove Theorem 4.98.

[145] [Rhi2001]

4.10 Disasters in Topology IV: The Baire Category Theorem

The applications of topology to analysis are usually manifested in the form of an "existence theorem" of some sort and the major share of the work in this direction is born, directly or indirectly, by two theorems: the Tychonoff theorem and the Baire category theorem.

S. Willard[146]

The category theorem, given by R. Baire in 1899 is one of the principal avenues through which applications of completeness are made in classical and functional analysis.

A. Wilansky[147]

The relevance of the Baire Category theorem to the fundamental metalogical principle of deductive completeness has long been known.

R. Goldblatt[148]

The above quotes indicate the usefulness of the Baire Category Theorem in different areas of mathematics. Like the Ascoli Theorem it comes in many different forms. We will restrict our attention to the more elementary forms given below.

Definition 4.99. *A topological space X is called* Baire *iff in X each countable intersection of dense, open subsets is dense.*

Definition 4.100. *The* ***Baire Category Theorem*** *states that all completely*[149] *metrizable spaces and all compact Hausdorff spaces are Baire.*

Disaster 4.101.
1. There may exist completely metrizable spaces that fail to be Baire.
2. There may exist compact Hausdorff spaces that fail to be Baire.

We will start by presenting classes of topological spaces that are Baire in **ZF** and continue by widening these classes by stepwise adding set theoretical conditions of increasing strength.

---

[146] [Wil70, p. 185]
[147] [Wila70, p. 178]
[148] [Gol85]
[149] A pseudometric space is called *complete* iff in X every Cauchy sequence converges. For a completeness concept based on Cauchy filters see Exercise E 4.

Theorem 4.102. [150] *Separable completely metrizable spaces are Baire.*

Proof. Let X be a topological space with a compatible complete metric d. Let (B_n) be a sequence of dense, open sets in X, let B be a non–empty open set in X, and let $\{x_n \mid n \in \mathbb{N}\}$ be dense in X. For each $n \in \mathbb{N}^+$ and each $x \in X$, define $S(x,n) = \{y \in X \mid d(x,y) < \frac{1}{n}\}$ and $T(x,y) = \{y \in X \mid d(x,y) \leq \frac{1}{n}\}$. Construct, via recursion, a sequence (y_n, r_n) of pairs, with $y_n \in X$ and $r_n > 0$ as follows:

$$\begin{aligned}
y_0 &= x_{\min\{m\in\mathbb{N} \mid x_m \in (B\cap B_0)\}} \\
r_0 &= \left(\min\{m \in \mathbb{N}^+ \mid T(y_0, m) \subseteq (B \cap B_0)\}\right)^{-1}. \\
y_{n+1} &= x_{\min\{m\in\mathbb{N} \mid x_m \in (S(y_n,r_n)\cap B_{n+1})\}}. \\
r_{n+1} &= \min\left\{\frac{1}{n+1}, \frac{1}{\min\{m\in\mathbb{N}^+ \mid T(y_{n+1},m)\subseteq(S(y_n,r_n)\cap B_{n+1})\}}\right\}.
\end{aligned}$$

Then (y_n) is a Cauchy–sequence, which thus converges to some point y. By construction,
$y \in (B \cap \bigcap_{n\in\mathbb{N}} B_n)$. Thus $\bigcap_{n\in\mathbb{N}} B_n$ is dense in X.

Theorem 4.103. [151] *Countably compact pseudometrizable spaces are Baire.*

Proof. Let X be a countably compact space with a compatible pseudometric d. Let (B_n), B, $S(x,n)$, and $T(x,n)$ be as in the previous proof. Construct, via recursion, a sequence of pairs (k_n, A_n), consisting of positive integers k_n and non–empty, open subsets A_n of X, as follows:

$$\begin{aligned}
k_0 &= \min\{m \in \mathbb{N}^+ \mid \exists x \in X \;\; T(x,m) \subseteq (B \cap B_0)\}. \\
A_0 &= \bigcup\{S(x,k_0) \mid T(x,k_0) \subseteq (B \cap B_0)\}. \\
k_{n+1} &= \min\{m \in \mathbb{N}^+ \mid \exists x \in X \;\; T(x,m) \subseteq (A_n \cap B_{n+1})\}. \\
A_{n+1} &= \bigcup\{S(x,k_{n+1}) \mid T(x,k_{n+1}) \subseteq (A_n \cap B_{n+1})\}.
\end{aligned}$$

Then $A_{n+1} \subseteq (A_n \cap B_n \cap B)$. Thus, by countable compactness of X,

$$\emptyset \neq \bigcap_{n\in\mathbb{N}} \mathrm{cl} A_n = \bigcap_{n\in\mathbb{N}} A_n \subseteq (B \cap \bigcap_{n\in\mathbb{N}} B_n)$$

. Consequently $\bigcap_{n\in\mathbb{N}} B_n$ is dense in X.

Theorem 4.104. [152] *Equivalent are:*

1. **CC**.
2. *Totally bounded, complete pseudometric spaces are Baire.*
3. *Second countable, complete pseudometric spaces are Baire.*

[150] [Bru83]
[151] [HeKe2000a]
[152] [BeHe98]

Proof. (**1**) ⇒ (**2**) By Proposition 3.26, (1) implies that every totally bounded, complete pseudometric space is compact. Thus (2) follows from Theorem 4.103.

(**2**) ⇒ (**3**) Immediate via Exercise E 3.

(**3**) ⇒ (**1**) Assume that **CC** fails. Then, by Theorem 2.12 resp. by Exercises to Section 2.2, E 4, there exists a sequence (X_n) of non–empty sets such that each sequence meets only finitely many X_n's. Let $\varphi\colon \mathbb{N} \to \{r \in \mathbb{Q} \mid 0 \leq r \leq 1\}$ be a bijection. Then the space (X, d), defined by:

$$X = \bigcup_{n \in \mathbb{N}} (X_n \times \{n\})$$

$$d\big((x,n),\ (y,m)\big) = \begin{cases} |\varphi(n) - \varphi(m)|, & \text{if } n \neq m \\ 0, & \text{if } n = m, \end{cases}$$

is a second countable complete pseudometric space that fails to be Baire, since each $B_n = X \setminus (X_n \times \{n\})$ is dense and open in (X, d), but $\bigcap_{n \in \mathbb{N}} B_n = \emptyset$.

Theorem 4.105. [153] *Equivalent are:*

1. **CC**.
2. *Countable products of compact pseudometric spaces are Baire.*
3. $X^{\mathbb{N}}$ *is Baire for each compact pseudometric space* X.

Proof. (**1**) ⇒ (**2**) Let $\big((X_n, d_n)\big)$ be a sequence of compact pseudometric spaces. Then $d\big((x_n),(y_n)\big) = \max\{\min\{2^{-n},\ d_n(x_n,y_n)\} \mid n \in \mathbb{N}\}$ defines a compatible pseudometric on the topological product space $X = \prod_{n \in \mathbb{N}} X_n$. Since each (X_n, d_n) is complete and totally bounded, so is (X, d). Thus, by (1) and Proposition 3.26, X is compact. Consequently Theorem 4.103 implies that X is Baire.

(**2**) ⇒ (**3**) Obvious.

(**3**) ⇒ (**1**) Let $(X_n)_{n \in \mathbb{N}}$ be a sequence of non–empty sets. Define a compact pseudometric space (X, d) by:

$$X = \{(0,0)\} \cup \bigcup_{n \in \mathbb{N}^+} (X_n \times \{n\})$$

$$d\big((x,n),(y,m)\big) = \begin{cases} 0, & \text{if } n+m = 0 \\ \frac{1}{n+m}, & \text{if } n+m \neq 0 \text{ and } n \cdot m = 0 \\ |\frac{1}{n} - \frac{1}{m}|, & \text{if } n \cdot m \neq 0. \end{cases}$$

In the product space $(X,d)^{\mathbb{N}}$, for each $n \in \mathbb{N}^+$, the set $B_n = \bigcup_{m \in \mathbb{N}^+} \pi_m^{-1}\big[X_n \times \{n\}\big]$ (where π_m denotes the m–th projection) is dense and open. Thus, by (3), there exists an element (d_n) in $\bigcap_{n \in \mathbb{N}^+} B_n$. Each d_n has the form $(y_{k(n)}, l(n))$.

---

[153] [HeKe2000a]

Thus $(d_n) \in B_m$ implies that there exists some $n \in \mathbb{N}^+$ with $(y_{k(n)}, l(n)) \in (X_m \times \{m\})$. Therefore the sequence $(y_{k(n)})_{n\in\mathbb{N}^+}$ meets each X_m. Consequently the sequence $(x_m)_{m\in\mathbb{N}^+}$, defined by $x_m = y_{k(\min\{n\in\mathbb{N}|y_{k(n)}\in X_m\})}$ is an element of $\prod_{m\in\mathbb{N}^+} X_m$.

Theorem 4.106. [154] *Equivalent are:*

1. **DC**.
2. *Complete pseudometric spaces are Baire.*
3. *a) Compact Hausdorff spaces are Bairc*
   and
   (b) Countable products of compact (Hausdorff) spaces are compact.
4. *Countable products of compact Hausdorff spaces are Baire.*
5. *Countable products of discrete spaces are Baire.*
6. *$X^{\mathbb{N}}$ is Baire, for each discrete space X.*
7. *$(\alpha X)^{\mathbb{N}}$ is Baire, for the 1–point–compactification αX of each discrete space X.*

Proof. **(1)** ⇒ **(2)** Let X be a complete pseudometric space with pseudometric d. For $x \in X$ and $r > 0$, define $S(x,r) = \{y \in X \mid d(x,y) < r\}$ and $T(x,r) = \{y \in X \mid d(x,y) \leq r\}$. Let (B_n) be a sequence of dense, open sets in X.

Consider:
$Y = \{(n,x,r) \in \mathbb{N} \times X \times \mathbb{R} \mid 0 < r < 2^{-n} \text{ and } T(x,r) \subseteq (B \cap \bigcap_{m\leq n} B_m)\}$.
Define a binary relation ϱ on Y by:

$$(n,x,r)\varrho(\bar{n},\bar{x},\bar{r}) \Leftrightarrow \big(n < \bar{n} \text{ and } T(\bar{x},\bar{r}) \subseteq S(x,r)\big).$$

Then, for each $y \in Y$, there exists some $\bar{y} \in Y$ with $y\varrho\bar{y}$. Thus, by (1), there exists a sequence (y_n) in Y with $y_n \varrho y_{n+1}$ for each n. This implies that, for $y_n = (m_n, x_n, r_n)$, the sequence (x_n) is Cauchy and thus converges to some point x in $\bigcap_{n\in\mathbb{N}} T(x_n, r_n) \subseteq (B \cap \bigcap_{m\in\mathbb{N}} B_m)$.

(1) ⇒ **(3)** (b) follows from Proposition 4.72. For (a), consider a compact Hausdorff space X, let (B_n) be a sequence of dense, open sets in X, and let B be a non–empty open set in X. Consider the set Y of all pairs (n, A), consisting of $n \in \mathbb{N}$ and non–empty, open subsets A of X such that cl$A$, the closure of A, is contained in $B \cap \bigcap_{m\leq n} B_m$. Define a binary relation ϱ on Y by:

$$(n,A)\varrho(\bar{n},\bar{A}) \Leftrightarrow (n < \bar{n} \text{ and } \mathrm{cl}\bar{A} \subseteq A).$$

Then, for each $y \in Y$, there exists $\bar{y} \in Y$ with $y\varrho\bar{y}$. Thus, by (1), there exists a sequence (y_n) in Y with $y_n \varrho y_{n+1}$ for each n. This implies that, for $y_n = (m_n, A_n)$, we obtain

[154] [Bla77], [Bru83], [HeKe2000a].

$$\cdots \mathrm{cl}A_{n+1} \subseteq A_n \subseteq \cdots \mathrm{cl}A_1 \subseteq A_0 \subseteq \mathrm{cl}A_0.$$

Thus there exists some point $x \in \bigcap_{n\in\mathbb{N}} \mathrm{cl}A_n \subseteq (B \cap \bigcap_{m\in\mathbb{N}} B_m)$.

(2) $\Rightarrow$ **(5)** Immediate, since countable products $X = \prod_{n\in\mathbb{N}} X_n$ of discrete spaces X_n are completely metrizable, e.g., by the metric

$$d\big((x_n),(y_n)\big) = \begin{cases} 0, & \text{if } (x_n) = (y_n) \\ 2^{-\min\{n\in\mathbb{N} \mid x_n \neq y_n\}}, & \text{otherwise.} \end{cases}$$

(5) $\Rightarrow$ **(6)** Obvious.

(6) $\Rightarrow$ **(1)** Let X be a non–empty set and ϱ a relation on X that satisfies:

$$\forall x \in X \quad \exists y \in X \quad x\varrho y.$$

Consider X as a discrete space. Then, by (6), the product space $Y = X^{\mathbb{N}}$ is Baire. For each $n \in \mathbb{N}$, define

$$B_m = \{(x_n) \in Y \mid \exists n \in \mathbb{N} \;\; x_m \varrho x_n\}.$$

Then the B_m's are dense and open in the non–empty space Y. Thus there exists some point (x_n) in $\bigcap_{m\in\mathbb{N}} B_m$. By construction:

$$\forall n \in \mathbb{N} \quad \exists m \in \mathbb{N} \quad x_n \varrho x_m.$$

Define, via recursion, a sequence $(\bar{x}_n)$ as follows:

$$\begin{aligned} \bar{x}_0 &= x_0 \\ \bar{x}_{n+1} &= x_{\min\{m\in\mathbb{N} \mid \bar{x}_n \varrho x_m\}} \end{aligned}.$$

Then $\bar{x}\varrho\bar{x}_{n+1}$ for each $n \in \mathbb{N}$.

(3) $\Rightarrow$ **(4)** $\Rightarrow$ **(7)** Obvious.

(7) $\Rightarrow$ **(1)** Follows precisely as in the above proof of the implication (6) $\Rightarrow$ (1).

Remark 4.107. By the above theorem the complete metric version of the Baire Category Theorem is equivalent to **DC**. Whether the compact Hausdorff version is not only implied by **DC**, but also equivalent to **DC** (resp., whether condition (3a) implies (3b) — and thus **DC**) remains an open question, cf. Exercises E 5. However, it is known that condition (3b) does not imply (3a) — and hence **DC**, as pointed out in Remark 4.73.

Further variants of the Baire Category Theorem can be obtained by modifying the concept of *completeness*. For a completeness concept, obtained by means of Cauchy filters instead of Cauchy sequences see Exercises E 4. Discussions of more complicated completeness concepts such as *Čech–completeness, pseudo–completeness, regular–closedness* or *pseudocompactness* are beyond the scope of this book. See in particular [Oxt61], [Gol85], [HeKe99].

Let us finally turn to another variant of the *compactness* concept, namely to ultrafilter–compactness, in order to demonstrate that even **DC** may not always suffice to obtain a suitable variant of the Baire Category Theorem.

Theorem 4.108. [155] *Equivalent are:*

1. **DC** *and* **WUF**.
2. **DC** *and* **WUF**(ℕ)
3. *Regular, ultrafilter–compact spaces are Baire.*

Proof. (**1**) ⇒ (**2**) Obvious.

(**2**) ⇒ (**1**) Let X be an infinite set. By Theorems 2.12 and 2.14, X is D–infinite, i.e., there exists an injection $f\colon \mathbb{N} \to X$: Let $\mathcal{U}$ be a free ultrafilter on ℕ. Then $\{V \subseteq X \mid f^{-1}[V] \in \mathcal{U}\}$ is a free ultrafilter on X.

(**2**) ⇒ (**3**) Let X be a regular ultrafilter–compact space, let (B_n) be a sequence of dense, open sets in X, and let B be a non–empty open set in X. Construct — as in the proof of the implication (1) ⇒ (3) of Theorem 4.106 — a sequence (A_n) of non–empty open sets in X with $\mathrm{cl}A_{n+1} \subseteq A_n$ for each $n \in \mathbb{N}$, and $\bigcap\limits_{n\in\mathbb{N}} A_n = \bigcap\limits_{n\in\mathbb{N}} \mathrm{cl}A_n \subseteq (B \cap \bigcap\limits_{m\in\mathbb{N}} B_m)$. By Theorem 2.12 there exists an element (a_n) in $\prod\limits_{n\in\mathbb{N}} A_n$. If $A = \{a_n \mid n \in \mathbb{N}\}$ is finite, there exists some $a \in A$ that is contained in infinitely many and thus in all A_m's, and consequently in $B \cap \bigcap\limits_{m\in\mathbb{N}} B_m$. Otherwise, A is countable, thus there exists a free ultrafilter $\mathcal{U}$ on A. Consequently $\mathcal{W} = \{V \subseteq X \mid (V \cap A) \in \mathcal{U}\}$ is a free ultrafilter on X, and hence converges to some point x. As a cluster point of $\mathcal{W}$, x belongs to $\mathrm{cl}A_n$ for each $n \in \mathbb{N}$, and thus to $B \cap \bigcap\limits_{m\in\mathbb{N}} B_m$.

(**3**) ⇒ (**2**) Let X be a discrete space, let αX be the 1–point–compactification of X, and let $Y = (\alpha X)^{\mathbb{N}}$ be the corresponding product. Then Y is a regular, ultrafilter–compact space, thus, by (3), Baire. Hence **DC** holds by Theorem 4.106. If there would be no free ultrafilter on ℕ, then the space ℚ would be a regular, ultrafilter–compact space which fails to be Baire — contradicting condition (3).

Note that the conditions **DC** and **WUF** are independent of each other. There exist models[156] of **ZF** that satisfy **PIT** and hence **WUF** but not **DC**; and vice versa, there exist models[157] of **ZF** that satisfy **DC**, but fail to satisfy even **WUF**(?).

Exercises to Section 4.10:

E 1. Show that finite intersections of dense, open sets are dense and open.

[155] [HeKe2000a]

[156] E.g., Cohen's First Model A4 (M1 in [HoRu98]).

[157] E.g., Pincus–Solovay's Model A6 (M27 in [HoRu98]).

E 2. [158] Show that, under **CC**, totally bounded pseudometric spaces are separable.

E 3. [159] Show that if a second countable space X is pseudometrizable there exists a compatible totally bounded pseudometric for X.

E 4. [160] A pseudometric space X is called *filter–complete* iff in X every Cauchy filter converges. Show that:
a) Every filter–complete pseudometric space is complete.
b) Every complete pseudometric space is filter–complete iff **CC** holds.
c) Second countable, filter–complete pseudometric spaces are Baire.

E 5. [161] Show the equivalence of:
a) Compact Hausdorff spaces are Baire.
b) **DMC**, the *Principle of Dependent Multiple Choices*, stating that for every non–empty set X and every relation ϱ on X satisfying

$$\forall x \in X \quad \exists y \in X \quad x \varrho y$$

there exists a sequence (F_n) of non–empty finite subsets F_n of X, satisfying

$$\forall_n \quad \forall x \in F_n \quad \exists y \in F_{n+1} \quad x \varrho y.$$

E 6. [162] Show that $\mathbf{CC}(\mathbb{R})$ implies that separable, compact Hausdorff spaces are Baire.

E 7. [163] Show the equivalence of:

a) **DC**.
b) Products of compact Hausdorff spaces are Baire.
c) Compact Hausdorff spaces and Cantor–cubes $\mathbf{2}^I$ are Baire.

E 8. [164] Show the equivalence of:
a) **CC**.
b) Sequentially compact pseudometric spaces are Baire.

E 9. [165] Show the equivalence of:
a) **CC**(fin).
b) Countable products of finite Hausdorff spaces are Baire.

---

158 [BeHe98]
159 [BeHe98]
160 [HeKe2000a]
161 [FoMo98]
162 [Ker2003]
163 [HeKe99], [HeKe99a].
164 [HeKe2000a]
165 [HeKe2000]

E 10. [166] Show that, for each $n \in \mathbb{N}$, the following conditions are equivalent:
a) $\mathbf{CC}(\leq n)$ [see Exercises to Section 3.1, E 1].
b) Countable products of Hausdorff spaces with at most $n+1$ points each are Baire.

E 11. [167] Show that the Baire property for Cantor–cubes $\mathbf{2}^I$ implies each of the following conditions:
a) $\mathcal{P}X$ is D–infinite, for each infinite set X.
b) $\mathbf{CC}(\text{fin})$.
c) There are no amorphous sets. [See Exercises to Section 4.1, E 11.]

4.11 Disasters in Graph Theory: Coloring Problems

I am sure that $a_2 > 4$ but cannot prove it.
P. Erdös[168]

To investigate coloring problems in graph theory we need to build up some terminology first:

Terminology 4.109. A *graph* is a pair (X, ϱ), consisting of a set X, whose elements are called *vertices*, and a symmetric, antireflexive binary relation ϱ on X (i.e., $x\varrho y \Rightarrow (y\varrho x \text{ and } x \neq y)$), whose elements (x, y) are called *edges*.

A *homomorphism* $f\colon (X, \varrho) \to (Y, \sigma)$ between graphs is a map $f\colon X \to Y$ satisfying

$$x\varrho y \Rightarrow f(x)\sigma f(y).$$

A graph (X, ϱ) is called a *subgraph* of a graph (Y, σ) provided that X is a subset of Y and $\varrho = \sigma_X$ is the restriction of σ to $X \times X$.

A graph (X, ϱ) is called *complete* provided that

$$\varrho = \{(x, y) \in X \times X \mid x \neq y\}.$$

$\mathbf{n}$ denotes the complete graph with $n = \{0, 1, \ldots, n-1\}$ as set of vertices; the elements of n being called *colors*.

An *n–coloration* of the graph G is a homomorphism $f\colon G \to \mathbf{n}$.

A graph G is called *n–colorable* provided there exists some n–coloration of G.

A graph (X, ϱ) is called *connected* provided that for any two distinct elements x and y of X there exists some tuple $(x_0, x_1, \ldots, x_n)$ with

$$x_0 = x, \; x_n = y, \text{ and } x_i \varrho x_{i+1} \text{ for each } i = 0, \ldots, n-1.$$

---

[166] [HeKe2000]
[167] [HeKe99a]
[168] [Erd80]. Here a_2 is the chromatic number $\chi(G)$ of the graph G, described in Exercise E 9.

The problem we are concerned with is whether a given graph is n–colorable, where n is some natural number. Obviously, $\mathbf{n+1}$ is not n–colorable. A necessary condition for the n–colorability of a graph G is that each of its finite subgraphs is n–colorable. In **ZFC** this condition is also sufficient:

Theorem 4.110. *In* **ZFC**, *for each n, the following conditions on a graph G are equivalent:*

1. *G is n–colorable.*
2. *Each finite subgraph of G is n–colorable.*

Proof. See Theorems 4.113 and 4.115 below.

The above result remains true in **ZF** only for $n = 0$ or $n = 1$.

Disaster 4.111. It may happen that every finite subgraph of some graph G is 2–colorable, but G fails to be n–colorable for any n.

Proof. Let $(X_n)_{n\in\mathbb{N}}$ be a sequence of 2–element sets with $\prod_{n\in\mathbb{N}} X_n = \emptyset$. Consider the graph $G = (X, \varrho)$ defined by $\begin{cases} X = \bigcup_{n\in\mathbb{N}} (X_n \times \{n\}) \\ \varrho = \{((x,n),(y,m)) \in X^2 \mid n=m \text{ and } x \neq y\} \end{cases}$.
Then every finite subgraph of G is 2–colorable, but G is n–colorable for no $n \in \mathbb{N}$.

For connected graphs and $n = 2$, Theorem 4.110 can be salvaged however:

Proposition 4.112. *For connected graphs G, the following conditions are equivalent:*

1. *G is 2–colorable.*
2. *Every finite subgraph of G is 2–colorable.*

Proof. Let every finite subgraph of the connected graph $G = (X, \varrho)$ be 2–colorable. If G is empty, nothing more need be said. Otherwise select an element a in x, and define, for each $x \in X$, $n(x)$ to be the smallest $n \in \mathbb{N}$ such that there exists some $(n+1)$–tuple $(x_0, x_1, \ldots, x_n)$ in X with $x_0 = a$, $x_n = x$, and $x_i \varrho x_{i+1}$ for $i = 0, \ldots, n-1$. Then the function $f \colon G \to \mathbf{2}$, defined by

$$f(x) = \begin{cases} 0, \text{ if } n(x) \text{ is even} \\ 1, \text{ if } n(x) \text{ is odd,} \end{cases}$$

is a 2–coloration of G. (Cf. Exercise E 1).

Combining the ideas that enter into the proofs of the results 4.111 and 4.112, we obtain:

Theorem 4.113. [169] *Equivalent are:*

1. *If every finite subgraph of a graph G is 2–colorable, then so is G.*
2. **AC(2)**.

Proof. **(1)** $\Rightarrow$ **(2)** Let $(X_i)_{i\in I}$ be a family of 2–element sets. Consider the graph $G = (X, \varrho)$, defined by $\begin{cases} X = \bigcup\limits_{i\in I} (X_i \times \{i\}) \\ \varrho = \{((x,i),(y,j)) \in X^2 \mid i = j \text{ and } x \neq y\}. \end{cases}$

Then every finite subgraph of G is 2–colorable. Thus, by (1), G itself is 2–colorable. Let $f\colon G \to \mathbf{2}$ be a 2–coloration of G. Then, for each $i \in I$, there exists precisely one element x_i of X_i with $f(x_i, i) = 0$. Thus $(x_i)_{i\in I} \in \prod\limits_{i\in I} X_i$.

(2) $\Rightarrow$ **(1)** Let $G = (X, \varrho)$ be a non–empty graph, such that all of its finite subgraphs are 2–colorable. Call a subset C of X a *component* of G provided that the subgraph (C, ϱ_C) of G, determined by C, is a maximal connected subgraph of G. Let I be the set of all components of G. Then for each C in I, the graph (C, ϱ_C) is connected and hence by Proposition 4.112, 2–colorable. Moreover, as can be seen easily (cf. Exercise E 2) for each C in I, the set X_C of all 2–colorations of (C, ϱ_C) contains precisely 2 elements. Thus, by (2), there exists a family $(f_C)_{C\in I}$ of 2–colorations $f_C\colon (C, \varrho_C) \to \mathbf{2}$. Consequently the function $f\colon G \to \mathbf{2}$, defined by

$$f(x) = f_C(x), \text{ if } x \in C$$

is a 2–coloration of G.

As can be seen easily (cf. Exercise E 3), the implication (1) $\Rightarrow$ (2) of the above Theorem 4.113 remains valid if the number 2 is replaced by any natural number $n \geq 3$. However, the inverse implication (2) $\Rightarrow$ (1) may fail.

Disaster 4.114. Even under **AC(3)**, there may exist graphs that fail to be 3–colorable even though all their finite subgraphs are 3–colorable.

Proof. This follows from the next theorem and the fact [170] that **AC(3)** does not imply **PIT**.

Theorem 4.115. [171] *Equivalent are:*

1. *If every finite subgraph of a graph G is 3–colorable, then so is G.*
2. **PIT**.

Proof. **(1)** $\Rightarrow$ **(2)** Let B be a Boolean algebra with $0 \neq 1$. We will construct a graph $G = (X, \varrho)$ such that

---

[169] [Myc61]

[170] Pincus' Model (M43 in [HoRu98]) satisfies **AC**(fin), but fails to satisfy **PIT**.

[171] [BrEr51], [Laeu71].

(a) Each finite subgraph of G is 3–colorable.
(b) If G is 3–colorable, then B has a maximal ideal.

This then, together with (1), will imply (2).
G will be constructed in 3 steps.

Step 1: $G_1 = (X_1, \varrho_1)$ with $X_1 = B$ and $\varrho_1 = \{(x, x^*) \mid x \in B\}$, where x^* is the complement of x in B.

Step 2: $G_2 = (X_2, \varrho_2)$ with $X_2 = X_1 \cup \{a\}$, where a is some element not contained in X_1 and $\varrho_2 = \varrho_1 \cup (\{a\} \times X_1) \cup (X_1 \times \{a\})$.

Step 3: Let P be the set of all subsets $\{x, y\}$ of B such that x and y are neither comparable nor dual to each other. Associate with each $\{x, y\} \in P$ a graph $G(x, y) = \big(A(x, y), \varrho(x, y)\big)$ as depicted:

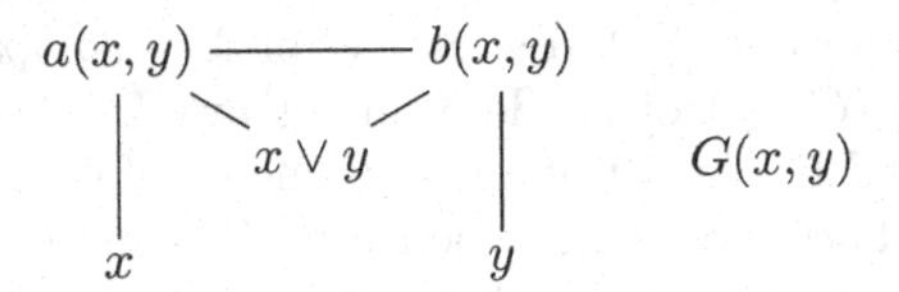

Define the graph $G = (X, \varrho)$ by:

$$X = X_2 \cup \bigcup_{\{x,y\} \in P} A(x, y)$$

$$\varrho = \varrho_2 \cup \bigcup_{\{x,y\} \in P} \varrho(x, y).$$

Step 1 guarantees that any 3–coloration f of G satisfies $f(x) \neq f(x^*)$ for each $x \in B$.

Step 2 guarantees that every 3–coloration f of G induces a 2–coloration of G_1 and that every 2–coloration of G_1 can be extended to a 3–coloration of G_2.

Step 3 guarantees that, for each $\{x, y\} \in P$, a map $f\colon \{x, y, x \vee y\} \to 3$ can be extended to a 3–coloration $\tilde{f}$ of $G(x, y)$ if and only if f satisfies the condition:

(*) If $f(x) = f(y)$ then f is constant.

Now we can prove (a) and (b):

Proof **of (a):** Let $H = (Z, \varrho_Z)$ be a finite subgraph of G. Then there exists a finite subalgebra K of B with $Z \subseteq \big(K \cup \{a\} \cup \bigcup_{\{x,y\} \in (P \cap \mathcal{P}K)} A(x, y)\big)$.

As a finite Boolean algebra with $0 \neq 1$, K has a maximal ideal I. Its complement $F = K \backslash I$ has the form $F = \{x \in K \mid x^* \in I\}$. We construct a 3–coloration $G\colon H \to \mathbf{3}$ in 3 steps:

Step 1: For $x \in (Z \cap X_1)$, define $G(x) = \begin{cases} 0, \text{ if } x \in I \\ 1, \text{ if } x \in F \end{cases}$.

Step 2: If $a \in Z$, define $G(a) = 2$.

Step 3: If $z \in (Z \cap A(x,y))$ for some $\{x,y\} \in (P \cap \mathcal{P}K)$, define $g(z) = \tilde{f}(z)$, where f is the restriction of g to the set $\{x, y, x \vee y\}$ and $\tilde{f}$ is a 3–coloration extension to $G(x,y)$, which exists since f satisfies the condition (*).
Proof **of (b):** Let $f\colon G \to \mathbf{3}$ be a 3–coloration of G. Assume w.l.o.g. that $f(a) = 2$ and $f(0) = 0$. Define $I = B \cap f^{-1}(0)$. Then $F = B \cap f^{-1}(1)$ is the complement $B \setminus I$ of I in B. Moreover, $\varrho_1 \subseteq \varrho$ implies that $F = \{x^* \mid x \in I\}$. Furthermore, for $\{x,y\} \in (P \cap \mathcal{P}I)$ we must have $(x \vee y) \in I$, since otherwise the restriction of f to $\{x, y, x \vee y\}$ would violate the above condition (*). Thus I is closed under joins $x \vee y$. Analogously F is closed under joins $x \vee y$. Hence, by Exercise E 5, I is a maximal ideal in B.

(2) $\Rightarrow$ **(1)** Let $G = (X, \varrho)$ be a graph such that each finite subgraph of G is 3–colorable. Let I be the set of all finite subgraphs of G. For each $K \in I$ let C_K be the set of all 3–colorations of K. Consider each C_K as a finite, discrete topological space. Then, by (2) and Exercises to Section 4.8, E 7, the product space $C = \prod_{K \in I} C_K$ is compact and non–empty. For every pair (L, R) of elements of I with $L \subseteq R$, the set $A(L,R) = \{(f_K) \in C \mid f|_L$ is the restriction f_R to $L\}$ is closed in C. Moreover, the sets $A(L,R)$ have the finite intersection property. Thus compactness of C implies that there exists some element $(f_K)_{K \in I}$ in the intersection of all the $A(L,R)$'s. For each $x \in X$ consider the subgraph $K_x = (\{x\}, \emptyset)$ of G, and define $f\colon X \to \mathbf{3}$ by $f(x) = f_{K_x}(x)$. Then $f\colon G \to \mathbf{3}$ is a 3–coloration of G.

Exercises to Section 4.11:

E 1. Define, for graphs G, the concept of *cycles* in G, and show that for connected graphs G the following conditions are equivalent:
(1) G is 2–colorable.
(2) G has no odd–numbered cycles.

E 2. Show that a connected graph G has either precisely two 2–colorations or none at all.

E 3. Show that, for every natural number $n \geq 2$, $\mathbf{AC}(n)$ is implied by the assumption that a graph is n colorable whenever all its finite subgraphs are so.

E 4. Let n be a natural number ≥ 4. Show the equivalence of:
(1) If every finite subgraph of a graph G is n–colorable, then so is G.
(2) **PIT**.
[Hint: Use Step 2 in the proof of Theorem 4.115.]

E 5. Let I be a subset of a Boolean algebra B, and $F = B \setminus I$. Show the equivalence of:
(1) I is a maximal ideal in B.

(2) I is a prime ideal in B.
(3) F is a maximal filter in B.
(4) F is a prime filter in B.
(5) The following conditions hold:
  (a) $0 \in I$.
  (b) $x \in I \Leftrightarrow x^* \in F$ (where x^* is the complement of x in B).
  (c) $(x \in I$ and $y \in I) \Rightarrow (x \vee y) \in I$.
  (d) $(x \in F$ and $y \in F) \Rightarrow (x \vee y) \in F$.
(6) The map $f\colon B \to \mathbf{2}$, where $\mathbf{2}$ is the 2–element Boolean algebra, defined by

$$f(x) = \begin{cases} 0, & \text{if } x \in I \\ 1, & \text{if } x \in F, \end{cases}$$

is a Boolean homomorphism.

E 6. Consider the category **Grph** of graphs and homomorphisms. Show that:
(1) **Grph** has products of non–empty families.
(2) **Grph** has no terminal object.
(3) **Grph** has coproducts (called sums).
(4) **Grph** has equalizers but not coequalizers.
(5) If $(G_i)_{i\in I}$ is a family of graphs, such that at least one member G_{i_0} is n–colorable, then so is their product $\prod_{i\in I} G_i$.
[Hint: If $f\colon G_{i_0} \to \mathbf{n}$ is an n–coloration of G_{i_0} and $\pi_{i_0}\colon \prod_{i\in I} G_i \to G_{i_0}$ is the i_0–th projection, then $f_0 \circ \pi_{i_0}\colon \prod_{i\in I} G_i \to n$ is an n–coloration.]
(6) For each $n \geq 2$, let $C_n = (n, \sigma_n)$ be defined by

$$m\sigma_n k \;\Leftrightarrow\; (|m-k| = 1 \text{ or } \{m,k\} = \{0, n-1\}).$$

Then:
  (a) For odd $n \geq 3$ the graph C_n is not 2–colorable.
  (b) Under $\mathbf{AC}(\mathbf{2})$ or $\mathbf{AC}(\mathbb{R})$, the product $\prod_{n\in\mathbb{N}} C_{2n+3}$ is 2–colorable.

(7) Equivalent are:
  (a) Sums of 2–colorable graphs are 2–colorable.
  (b) $\mathbf{AC}(\mathbf{2})$.

(8) Let $f\colon G \to H$ be a homomorphism. If H is n–colorable, then so is G.
(9) Every graph $G = (X, \varrho)$ is a quotient of a 2–colorable graph $H = (Y, \sigma)$, i.e., there exists a surjective homomorphism $f\colon H \to G$ such that for any pair $(x_1, x_2) \in \varrho$ there exists some pair $(y_1, y_2) \in \sigma$ with $f(y_1) = x_1$ and $f(y_2) = x_2$.
[Hint: Define $Y = X \uplus (\varrho \times \{0,1\})$,

$$\sigma = \Big\{ \Big(((x,y),i), ((x,y),j) \Big) \mid (x,y) \in \varrho \text{ and } \{i,j\} = \{0,1\} \Big\},$$

$$f\colon Y \to X \ \text{ by } f(a) = \begin{cases} a, \text{ if } a \in X \\ x, \text{ if } a = \big((x,y),0\big) \\ y, \text{ if } a = \big((x,y),1\big), \end{cases}$$

and $g\colon H \to \mathbf{2}$ by $g(a) = \begin{cases} 0, \text{ if } a \in X \\ i, \text{ if } a = \big((x,y),i\big). \end{cases}$]

E 7. For each set X, embed the graph $G(X) = (\mathcal{P}X, \{(A, X\backslash A) \mid A \subseteq X\})$ into a graph $H(X)$ such that for each filter $\mathcal{F}$ on X the following conditions are equivalent:
(1) There exists a 3–coloration of $H(X)$ that is constant on $\mathcal{F}$.
(2) $\mathcal{F}$ can be enlarged to an ultrafilter on X.

E 8. Show that a graph $G = (X, \varrho)$ is
(1) 0–colorable iff $X = \emptyset$.
(2) 1–colorable iff $\varrho = \emptyset$.

E 9. [172] For any graph G, that is n–colorable for some n, its *chromatic number* $\chi(G)$ is the smallest n for which G is n–colorable.
Consider the graph $G = (\mathbb{R}^2, \varrho)$, defined by $x\varrho y$ iff the distance between x and y is one. Show that:
(1) $4 \leq \chi(G)$.
(2) $\chi(G) \leq 7$.
(3) If $\chi(G) = 4$ and **DC** holds, then there exist non–Lebesgue measurable subsets of $\mathbb{R}$.
Further:
(4) Investigate whether $\chi(G)$ depends on some choice–principle.
[Hints: For (1) consider the Moser Spindle

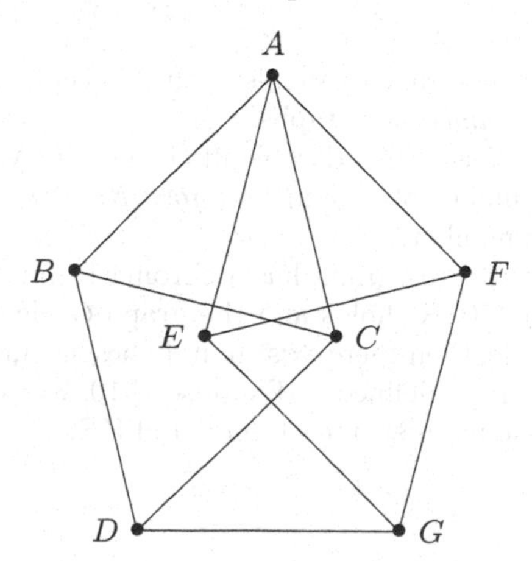

as a finite subgraph of G. For (2) tile the plane by regular hexagons of suitable size].

172 [Soi2003], [Fal81].

E 10. [173] Consider the Shelah–Soifer graph $G = (\mathbb{R}, \varrho)$, defined by

$x \varrho y \quad \Leftrightarrow \quad \exists r \in \mathbb{Q} \;\; \exists \xi \in \{\sqrt{2}, -\sqrt{2}\} \quad y = x + r + \xi.$

Show that:

(1) G has no odd–numbered cycles.
(2) Under $\mathbf{AC}(\mathbb{R})$ or $\mathbf{AC}(2)$, G is 2–colorable.
(3) If $f\colon G \to \mathbf{2}$ is a 2–coloration of G, then the sets $f^{-1}(0)$ and $f^{-1}(1)$ are non–Lebesgue–measurable.
(4) If $f\colon G \to \mathbf{n}$ is an n–coloration of G, then at least one of the coloring sets $f^{-1}(i)$ is non–Lebesgue–measurable.

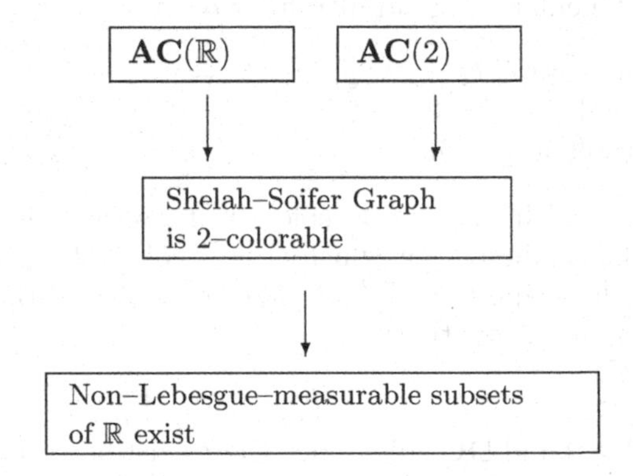

[Hint: For (3): Use the Exercises to Section 5.1, E 14 and the fact that the sets $f^{-1}(0)$ and $f^{-1}(1)$ are congruent to each other via arbitrary small shifts.]

E 11. [174]

(1) Define, for arbitrary cardinals c, the concepts of *c–colorability* and of *chromatic number* for graphs.
(2) Show that X is well–orderable iff the graph $(X \times \aleph, \varrho)$, where $\aleph$ is the Hartogs–number of X and $(x, \alpha)\varrho(y, \beta)$ iff $x \neq y$ and $\alpha \neq \beta$, has a chromatic number.
(3) Show that it every graph has a chromatic number iff **AC** holds.
(4) Show that $\mathbf{CC}(\mathbb{R})$ holds and the graph G, defined in E 10. above, is $\aleph_0$–colorable, then there exist non–Lebesgue–measurable subsets of $\mathbb{R}$.
(5) Is the graph G, defined in Exercise E 10, $\aleph_1$–colorable? Does G have a chromatic number in each model of **ZF**?

[173] [ShSo2003], [HeRh2005].
[174] [GaKo91], [ShSo2003], [HeRh2005].

5

Disasters with Choice

> *As is well known there are just as many (Lebesgue–) measurable sets as there are non–measurable ones: namely* $2^{\aleph}$ [where $\aleph = |\mathbb{R}|$]. *This is peculiar since we are used to the fact that in real analysis the pathologies predominate.*
>
> J. von Neumann[1]

However, as von Neumann points out, the above anomaly is only apparent — caused by the fact that there are so many sets of measure zero (namely $2^{\aleph}$). If sets A and B are called equivalent provided that their symmetric difference $A \triangle B$ is of measure zero, the *normal pathology of real analysis* in the **ZFC**–setting is restored: there are still $2^{\aleph}$ equivalence classes consisting of non–measurable sets (equivalently: containing a non–measurable set) but only $\aleph$ equivalence classes consisting of measurable sets (equivalently: containing a measurable set).

Observe further that not even complex analysis is devoid of pathologies. See, e.g., Exercises to Section 5.1, E 7.

5.1 Disasters in Elementary Analysis

> *Logic sometimes breeds monsters. For half a century there has been springing up a host of weird functions which seems to strive to have as little resemblance as possible to honest functions that are of some use ... They are invented on purpose to show our ancestor's reasonings at fault, and we shall never get anything more out of them.*
>
> H. Poincaré[2]

---

1 [vNeu29, p. 86]

2 *Mathematical definitions and education* (1906). Taken from [Fef2000].

Though the Axiom of Choice is responsible for many beautiful results, it is equally responsible for the existence of several dreadful monstrosities — unwelcome and unneeded.

Definition 5.1. *The equation* $f(x+y) = f(x) + f(y)$ *is called the* Cauchy–equation.

Consider a function $f\colon \mathbb{R} \to \mathbb{R}$ that satisfies the Cauchy–equation for all real x and y. Then it is easily seen that

- $f(r \cdot x) = r \cdot f(x)$ for all rational r and real x, i.e., f is $\mathbb{Q}$*–linear.*

In particular:

- $f(r) = f(1) \cdot r$ for all rational r.

So continuity of f would imply:

- $f(x) = f(1) \cdot x$ for all $x \in \mathbb{R}$.

Are there solutions of the Cauchy–equation that fail to be continuous? None has ever been constructed and in **ZF** none will ever be, since there are **ZF**–models without non–continuous solutions of the Cauchy–equation[3]. However the Axiom of Choice guarantees the existence of such monsters[4]; even worse, under **AC** there are far more undesirable solutions of the Cauchy–equation than there are desirable ones:

Disaster 5.2. In **ZFC** there are

1. $2^{\aleph_0}$ continuous solutions $f\colon \mathbb{R} \to \mathbb{R}$, and
2. $2^{(2^{\aleph_0})}$ non–continuous solutions $f\colon \mathbb{R} \to \mathbb{R}$ of the Cauchy–equation.

Proof. (1) For each $r \in \mathbb{R}$ the function $f\colon \mathbb{R} \to \mathbb{R}$, defined by $f(x) = r \cdot x$ is a continuous solution of the equation (1). There are no others, as noted above.

(2) In **ZFC**, $\mathbb{R}$, considered as a vector space over $\mathbb{Q}$, has a basis B, also called a *Hamel basis.* Moreover, a simple computation shows that $|B| = 2^{\aleph_0}$. Since any map $B \to \mathbb{R}$ can be extended uniquely to a linear map $\mathbb{R} \to \mathbb{R}$, there are precisely

$$|\mathbb{R}^B| = |\mathbb{R}|^{|B|} = (2^{\aleph_0})^{2^{\aleph_0}} = 2^{\aleph_0 \cdot 2^{\aleph_0}} = 2^{(2^{\aleph_0})}$$

solutions to be Cauchy–equation. Since only $2^{\aleph_0}$ of these are continuous, there are $2^{(2^{\aleph_0})}$ non–continuous ones.

Moreover, the non–continuous solutions of the Cauchy–equation have rather strange and unwanted features.

---

[3] E.g., Shelah's Second Model A2 (M38 in [HoRu98]).

[4] Observe, however that, although **AC** is the culprit in the present case, similar monsters can be constructed in **ZF**. See Exercise E 9.

Definition 5.3. *Non–continuous solutions of the Cauchy–equation are called* ugly.

Theorem 5.4. [5] *If* $f\colon \mathbb{R} \to \mathbb{R}$ *is ugly, then its graph* $G(f) = \{(x, f(x)) \mid x \in \mathbb{R}\}$ *is dense in* $\mathbb{R}^2$.

Proof. Let (x, y) be an element of $\mathbb{R}^2$ and let U be a neighborhood of (x, y) in $\mathbb{R}^2$. Since f is ugly there exist real numbers $a \neq 0$ and $b \neq 0$ such that the quotients $\alpha = \frac{f(a)}{a}$ and $\beta = \frac{f(b)}{b}$ are different. Consequently, $u = (a, f(a))$ and $v = (b, f(b))$ are linearly independent vectors in the real vector spacc $\mathbb{R}^2$, and thus form a basis of $\mathbb{R}^2$. Consequently, there exist real numbers[6] p and q with $(x, y) = p \cdot u + q \cdot v$. Since $\mathbb{Q}^2$ is dense in $\mathbb{R}^2$ and the expression $p \cdot u + q \cdot v$ depends continuously on p and q, there exist rational numbers $\bar{p}$ and $\bar{q}$ with $(\bar{p} \cdot u + \bar{q} \cdot v) \in U$. However

$$\bar{p} \cdot u + \bar{q} \cdot v = (\bar{p} \cdot a + \bar{q} \cdot b,\ \bar{p} \cdot f(a) + \bar{q} \cdot f(b)) = (\bar{p} \cdot a + \bar{q} \cdot b,\ f(\bar{p} \cdot a + \bar{q} \cdot b)).$$

Thus $(\bar{p} \cdot u + \bar{q} \cdot v) \in (U \cap G(f))$.

Theorem 5.5. [7] *Ugly functions are non–measurable.*

Proof. Let f be ugly. Assume[8] w.l.o.g. that there exist real numbers $a \neq 0$ and $b \neq 0$ with $f(a) = 1$ and $f(b) = 0$. For $n \in \mathbb{Z}$, define $A_n = f^{-1}[n, n+1)$ and choose $q_n \in \mathbb{Q}$ with $|n \cdot a - q_n \cdot b| < \frac{1}{2}$. Define $B_0 = A_0 \cap [-\frac{1}{2}, \frac{3}{2}]$ and

$$B_n = B_0 + n \cdot a - q_n \cdot b = \{x + n \cdot a - q_n \cdot b \mid x \in B_0\} \text{ for } n \neq 0.$$

Then $x \in (A_n \cap [0,1])$ implies that $y = x - (n \cdot a - q_n \cdot b) \in (A_0 \cap [-\frac{1}{2}, \frac{3}{2}])$, i.e., $y \in B_0$, and thus $x = y + (n \cdot a - q_n \cdot b) \in B_n$. Consequently

$$(A_n \cap [0,1]) \subseteq B_n \subseteq [-1, 2].$$

Thus:

$$[0,1] = [0,1] \cap \bigcup_{n \in \mathbb{Z}} A_n = \bigcup_{n \in \mathbb{N}} ([0,1] \cap A_n) \subseteq \bigcup_{n \in \mathbb{Z}} B_n \subseteq [-1, 2].$$

This implies that B_0 (and hence A_0) are non–measurable, since otherwise the B_n's, being pairwise disjoint and pairwise congruent, would have the same measure $\mu(B_n) = \mu(B_0)$ and thus[9]

---

5 [Ham05]

6 namely $p = \frac{y - \beta x}{a \cdot (\alpha - \beta)}$ and $q = \frac{\alpha \cdot x - y}{b \cdot (\alpha - \beta)}$.

7 [Sie20], [Bana20], [Kac36/37], [AlOr45], [Halp51].

8 If necessary, choose real numbers $a \neq 0$ and $b \neq 0$ with $\frac{f(a)}{a} \neq \frac{f(b)}{b}$ and replace f by the function g, defined by $g(x) = (b \cdot f(a) - a \cdot f(b))^{-1} \cdot (b \cdot f(x) - f(b) \cdot x)$.

9 Here we use σ–additivity of Lebesgue–measure (which requires some choice principle, cf. Exercise E 13.). However, our use of choice principles can be avoided. See Exercise E 14.

$$1 = \mu([0,1]) = \sum_{n\in\mathbb{Z}} \mu(B_0) \leq \mu([-1,2]) = 3,$$

which is impossible. Consequently f is not measurable.

This result guarantees the existence of some further monsters:

Disaster 5.6. In **ZFC** there are

1. $2^{(2^{\aleph_0})}$ non–measurable functions f that satisfy the Cauchy–equation,
2. $2^{(2^{\aleph_0})}$ non–measurable subsets of $\mathbb{R}$.

Proof. (**1**) Immediate from Disaster 5.2 and Theorem 5.5.

(**2**) The existence of non–measurable subsets of $\mathbb{R}$ follows immediately from (1). Their number is computed easily via the following "construction" of the *Vitali monsters.*

5.7. The Vitali Monsters[10] V

Let ϱ be the equivalence relation on $\mathbb{R}$, defined by

$$x\varrho y \Leftrightarrow (x-y)\in\mathbb{Q}.$$

Then each of the equivalence classes w.r.t. ϱ is dense in $\mathbb{R}$, thus meets the interval $[0,1]$. By $\mathbf{AC}(\mathbb{R})$, there exists a subset V of $[0,1]$ that contains precisely one element of each of these equivalence classes. The set $I = \mathbb{Q}\cap[-1,1]$ is countable. For each $r\in I$ define $V_r = \{v+r \mid v\in V\}$. Then $A = \bigcup_{r\in I} V_r$ is a countable union of pairwise disjoint sets satisfying $[0,1]\subseteq A\subseteq[-1,2]$. If V would be measurable then each of the V_r's would be measurable and would have the same measure as V. Thus A would be measurable and its measure would be 0, in case V would have measure 0, and ∞, otherwise. The former is not possible, since $[0,1]\subseteq A$; and the latter is not possible, since $A\subseteq[-1,2]$. Consequently, V is not measurable. Since there are precisely $2^{\aleph_0}$ equivalence classes w.r.t. ϱ, there exist precisely $\aleph_0^{(2^{\aleph_0})} = 2^{(2^{\aleph_0})}$ Vitali Monsters.

Here follow some different monster productions in **ZFC**:

5.8. The Bernstein Monsters[11] **B**

Since the space of reals has a countable base, it has precisely $2^{\aleph_0}$ open sets, and thus precisely $2^{\aleph_0}$ closed sets. By means of $\mathbf{AC}(\mathbb{R})$ it is easily deduced that each uncountable closed subset of $\mathbb{R}$ contains at least two complete accumulation points, thus a Cantor set, and thus has cardinality $2^{\aleph_0}$. Let $\mathfrak{A}$ be the set of all uncountable closed subsets of $[0,1]$. Then $|\mathfrak{A}| = |\mathbb{R}| = 2^{\aleph_0}$. By $\mathbf{AC}(\mathbb{R})$, $2^{\aleph_0} = \aleph$ for some Aleph $\aleph$. Thus $\mathfrak{A}$ can be expressed in the form $\mathfrak{A} = \{A_\alpha \mid \alpha < \aleph\}$. Let $f\colon \mathcal{P}_0(\mathbb{R}) \to \mathbb{R}$ be a map that satisfies $f(X)\in X$

[10] [Vit05]

[11] [Ber08]. Cf. also [Oxt80].

for each non–empty subset X of $\mathbb{R}$. Construct, via transfinite recursion, a transfinite sequence $(x_\alpha, y_\alpha)_{\alpha<\aleph}$ in $[0,1]^2$ as follows:

Assume that $\alpha < \aleph$, and that $(x_\beta, y_\beta)_{\beta<\alpha}$ have been defined already. Then the set $F = \{x_\beta \mid \beta < \alpha\} \cup \{y_\beta \mid \beta < \alpha\}$ has cardinality less than $\aleph$. Thus $A_\alpha \setminus F$ is not empty, and is in fact of cardinality $\aleph$. Define:

$$\begin{aligned} x_\alpha &= f[A_\alpha \setminus F] \\ y_\alpha &= f[A_\alpha \setminus (F \cup \{x_\alpha\})]. \end{aligned}$$

Then the sets $B = \{x_\alpha \mid \alpha < \aleph\}$ and $C = [0,1] \setminus B$ form a partition of $[0,1]$, and each member of $\mathfrak{A}$ meets each of the sets B and C. In other words: no member of $\mathfrak{A}$ is contained in B or in C. Hence every closed subset of B resp. of C is at most countable, and thus has Lebesgue–measure zero. Thus, if B would be Lebesgue measurable, then so would be C and both would have Lebesgue measure zero, which is impossible since their union $B \cup C = [0,1]$ has Lebesgue measure one.

5.9. The Sierpiński Monsters[12] S

Consider a free ultrafilter $\mathcal{U}$ on $\mathbb{N}$. Then the following hold:

1. $A \in \mathcal{U} \Leftrightarrow (\mathbb{N} \setminus A) \notin \mathcal{U}$ for each $A \subseteq \mathbb{N}$.
2. $(A \in \mathcal{U}$ and $F \subseteq \mathbb{N}$ finite$) \Rightarrow (A \Delta F) \in \mathcal{U}$.[13]

For each $A \subseteq \mathbb{N}$, let $\chi_A \colon \mathbb{N} \to \{0,1\}$ be the characteristic function of A, i.e., $A = \chi_A^{-1}(1)$. Consider $X = \{0,1\}^{\mathbb{N}}$. Define, for $x = (x_n)_{n\in\mathbb{N}}$, the element $x^* = (1 - x_n)_{n\in\mathbb{N}}$. With $S = \{\chi_U \mid U \in \mathcal{U}\}$ the above conditions (1) and (2) translate into

(1*) $X \setminus S = \{x^* \mid x \in S\}$,

(2*) $\left\{ \begin{array}{l} (x_n)_{n\in\mathbb{N}} \in S \\ (y_n)_{n\in\mathbb{N}} \in X \\ \{n \mid x_n \neq y_n\} \text{ finite} \end{array} \right\} \Rightarrow (y_n)_{n\in\mathbb{N}} \in S.$

Consider the normed product measure μ on X, invariant under the operation $*$. If S would be μ–measurable, then so would be $X \setminus S$ and, by (1*), $\mu(S) = \mu(X \setminus S)$ would hold. Thus $\mu(S) = \mu(X \setminus S) = \frac{1}{2}$.

But by (2*), Kolmogoroff's Zero–One–Law[14] would imply that either $\mu(S) = 0$ or $\mu(S) = 1$; a contradiction. From these observations it follows that $g[S]$ is not Lebesgue measurable, where $g \colon X \to [0,1]$ is defined by $g(x_n) = \sum\limits_{n\in\mathbb{N}} \frac{x_n}{2^{n+1}}$.

The following diagram illustrates which choice principles suffice to create certain monsters. For details not covered by the main text see the exercises.

---

12 [Sie38]

13 $A \Delta F = (A \cup F) \setminus (A \cap F) = (A \setminus F) \cup (F \setminus A)$

14 cf. [HewSt69, Theorem 22.21].

In Shelah's **ZF**–model[15] none of these principles holds, since it contains none of the above monsters — even though it satisfies a "reasonable" choice principle, namely **DC**, the Principle of Dependent Choices.

Diagram 5.10.

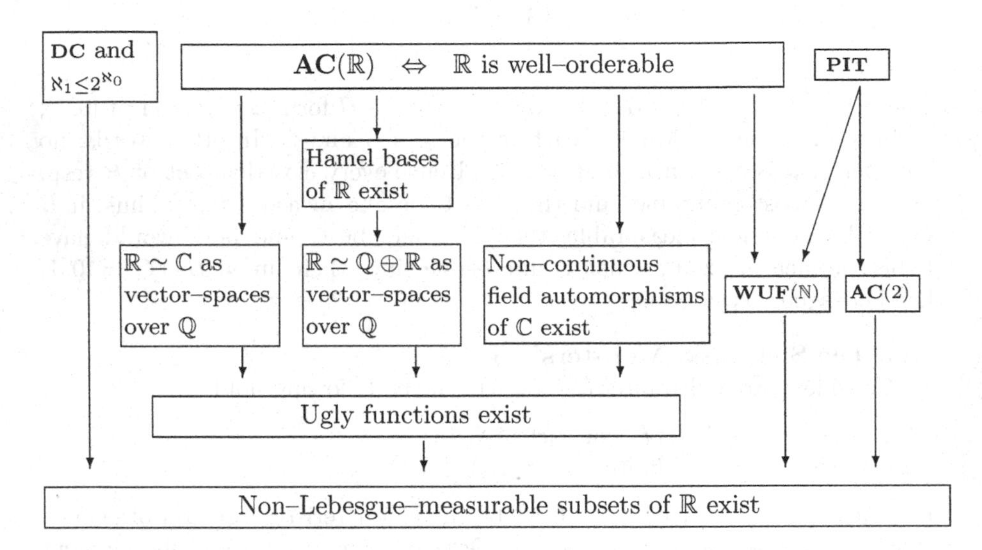

Besides monstrosities the Axiom of Choice produces also some harmless but somewhat bizarre curiosities. Here an example:

Curiosity 5.11. [16] In **ZFC** there exists a subset of the plane that meets every straight line, lying in the plane, in exactly two points.

Proof. The set L of all straight lines l in the plane P has cardinality $2^{\aleph_0}$, and so has each of the lines l. By **AC**, $2^{\aleph_0}$ is an aleph $\aleph$. So L can be written in the form $L = \{l_\alpha \mid \alpha < \aleph\}$. Also by **AC**, there exists a function $f\colon \mathcal{P}_0(P) \to P$ with $f(A) \in A$ for each non–empty subset A of P. Via transfinite recursion we define a transfinite sequence $(C_\alpha)_{\alpha<\aleph}$ of subsets of P such that $|C_\alpha| < \aleph$ and $|C_\alpha \cap l_\beta| \leq 2$ for each $\alpha < \aleph$ and $\beta < \aleph$.

Consider $\alpha < \aleph$, assume that C_β is defined for all $\beta < \alpha$. Let α' be the smallest of all $\gamma < \aleph$ such that $|l_\gamma \cap \bigcup_{\beta<\alpha} C_\beta| \leq 1$. Since $|\bigcup_{\gamma<\alpha'} (l_\gamma \cap l_{\alpha'})| < \aleph$, the set $B_\alpha = l_{\alpha'} \setminus (\bigcup_{\gamma<\alpha'} l_\gamma \cup \bigcup_{\beta<\alpha} C_\beta)$ is non–empty. Define $C_\alpha = \bigcup_{\beta<\alpha} C_\beta \cup f(B_\alpha)$.

Then $C = \bigcup_{\alpha<\aleph} C_\alpha$ meets every $l \in L$ in exactly 2 points.

---

[15] A2 (M38 in [HoRu98])

[16] S. Mazurkiewicz 1914; taken from [Sie58].

Exercises to Section 5.1:

E 1. [17] Show that **AC(2)** implies that there exist non–Lebesgue–measurable subsets of $\mathbb{R}$.
[Hint: With the notation of 5.9 define an equivalence relation ϱ on X by $(x_n)\varrho(y_n) \Leftrightarrow \{n \in \mathbb{N} \mid x_n \neq y_n\}$ finite. Let $[x]_\varrho$ be the equivalence class of $x \in X$. Consider the set $M = \{\{[x]_\varrho, [x^*]_\varrho\} \mid x \in X\}$ of 2-element sets. By **AC(2)**, there exists a set C of equivalence classes that contains exactly one element from each of the sets $\{[x]_\varrho, [x^*]_\varrho\}$. Then $S = \bigcup_{[x]_\varrho \in C} [x]_\varrho$ satisfies (1*) and (2*). Proceed as in 5.9.
Alternatively, assume that there exists a set S that contains for each function $f\colon \mathbb{R} \to \mathbb{R}$ exactly one member of the set $\{f, -f\}$. Define for each irrational number p a function

$$f_p\colon \mathbb{R} \to \mathbb{R} \text{ by } f_p(x) = \begin{cases} 1, \text{ if } (x-p) \in \mathbb{Q} \\ -1, \text{ if } (x+p) \in \mathbb{Q} \\ 0, \text{ otherwise} \end{cases}$$

and show that $X = \{p \in (\mathbb{R} \setminus \mathbb{Q}) \mid f_p \in S\}$ is non Lebesgue–measurable.]

E 2. [18] Show that **AC($\mathbb{R}$)** implies that $\mathbb{R} \simeq \mathbb{C}$, i.e., the sets $\mathbb{R}$ of all real and $\mathbb{C}$ of all complex numbers, considered as additive groups (equivalently as vector spaces over $\mathbb{Q}$), are isomorphic.
[Hint: Observe that $\mathbb{C} \simeq \mathbb{R} \oplus \mathbb{R}$ and that, by **AC($\mathbb{R}$)**, for any Hamel basis $(b_i)_{i \in I}$ we have $|I| = 2^{\aleph_0} = 2^{\aleph_0} + 2^{\aleph_0} = |I| + |I|$.]

E 3. Show that for any Hamel basis $(b_i)_I$ the indexing set I is D–infinite.
[Hint: Since the algebraic reals are countable, there exist a transcendental real t. Thus $(t^n)_{n \in \mathbb{N}}$ is linearly independent. Consider $t^n = \sum_{i \in I} \alpha(n,i) b_i$ with
$F_n = \{i \in I \mid \alpha(n,i) \neq 0\}$ finite. Then $\bigcup_{n \in \mathbb{N}} F_n$ is infinite. Define
$f\colon \mathbb{N} \to \mathbb{N}$ by $f(n) = \min\{k \in \mathbb{N} \mid (F_k \setminus \bigcup_{i<f(n)} F_i) \neq \emptyset\}$ and
$g\colon \mathbb{N} \to I$ by $g(n) = \min(F_{f(n)} \setminus \bigcup_{i<f(n)} F_i)$. Then g is injective.]

E 4. [19] Show that the existence of a Hamel basis implies that $\mathbb{R}$ and $\mathbb{R} \oplus \mathbb{Q}$ are isomorphic as vector spaces over $\mathbb{Q}$ (equivalently: as additive groups), i.e., $\mathbb{R} \simeq \mathbb{R} \oplus \mathbb{Q}$.
[Hint: Apply E 3. above.]

E 5. Show that each of the conditions
(1) $\mathbb{R} \simeq \mathbb{C}$,
(2) $\mathbb{R} \simeq \mathbb{R} \oplus \mathbb{Q}$,

[17] [Sie27], [Oxt80].
[18] [Ash75]
[19] [Ash75]

implies the existence of ugly functions.
[Hint:
(1) Let $h\colon \mathbb{R} \to \mathbb{C}$ be an isomorphism. Define $f\colon \mathbb{C} \to \mathbb{R}$ by $f(x+iy) = x$. Consider $f \circ h$.
(2) Let $h\colon \mathbb{R} \to \mathbb{R} \oplus \mathbb{Q}$ be an isomorphism. Define $f\colon \mathbb{R} \oplus \mathbb{Q} \to \mathbb{R}$ by $f(x, q) = x$. Consider $f \circ h$.]

E 6. Discuss whether it is
(1) desirable,
(2) undesirable
that there exists a Hamel basis for the reals. Present arguments for (a) and for (b).

E 7. [20] Consider $\mathbb{R}$ and $\mathbb{C}$ as fields. Show that:
(1) There exists precisely one automorphism of $\mathbb{R}$.
(2) There exist precisely two continuous automorphisms of $\mathbb{C}$.
(3) Under **AC**($\mathbb{R}$) there exist $2^{2^{\aleph_0}}$ non–continuous automorphisms of $\mathbb{C}$.
(4) If there exist non–continuous automorphisms of $\mathbb{C}$, there exist non–Lebesgue–measurable subsets of $\mathbb{R}$.
[Hint for (1): Observe that, in view of the equation $f(x^2) = \big(f(x)\big)^2$, any automorphism of $\mathbb{R}$ preserves order. For (4): Observe that whenever f a is non–continuous automorphism of $\mathbb{C}$, then the map $g\colon \mathbb{R} \to \mathbb{R}$, defined by[21] $f(x) = \Im\big(f(x)\big)$ is ugly.]

E 8. Show that in **ZFC** there exist isomorphisms $f\colon \mathbb{R} \to \mathbb{R}$ of the vector space $\mathbb{R}$ over $\mathbb{Q}$ into itself whose graphs are dense in $\mathbb{R}^2$.

E 9. Define $V = \{p + q \cdot \pi \mid (p, q) \in \mathbb{Q}^2\}$,
$j\colon \mathbb{Q}^2 \to V$ by $j(p, q) = p + q \cdot \pi$,
$f\colon \mathbb{Q}^2 \to \mathbb{Q}^2$ by $f(p, q) = (2p, q)$,
$\bar{f}\colon V \to V$ by $\bar{f}(p + q \cdot \pi) = 2p + q \cdot \pi$. Show that:

(1) The diagram
$$\begin{array}{ccc} \mathbb{Q}^2 & \xrightarrow{\;j\;} & V \\ {\scriptstyle f}\big\downarrow & & \big\downarrow{\scriptstyle \bar{f}} \\ \mathbb{Q}^2 & \xrightarrow[\;j\;]{} & V \end{array}$$
commutes.
(2) j, f and $\bar{f}$ are $\mathbb{Q}$–linear isomorphims.
(3) j is continuous and f is a homeomorphism.
(4) The graph of $\bar{f}$, $G(\bar{f}) = \{(v, \bar{f}(v) \mid v \in V\}$ is dense in V^2 hence also in $\mathbb{R}^2$.
[Hint: Cf. the proof of Theorem 5.4.]

E 10. Show that in **ZF** the vector space $\mathbb{Q}^{(\mathbb{N})}$ can be embedded as a $\mathbb{Q}$–linear subspace of $\mathbb{R}$. [Hint: Cf. the hint for E 3.)

---
20 [Kes51], [Sie58, p. 443].
21 Here $\Im$ is defined by $\Im(a + b \cdot i) = b$.

E 11. Show that (1) $\Rightarrow$ (2) $\Rightarrow$ (3):
(1) $\mathbf{AC}(\mathbb{R})$.
(2) Each linear subspace of the vector space $\mathbb{Q}^{(2^{\aleph_0})}$ has a linear complement.
(3) $\prod_{i \in I} X_i \neq \emptyset$ for each family $(X_i)_{i \in I}$ of pairwise disjoint, non–empty subsets of $\mathbb{R}$.

E 12. [22] Show that, under $\mathbf{DC}$ and $\aleph_1 \leq 2^{\aleph_0}$, there exist non–Lebesgue–measurable subsets of $\mathbb{R}$.

E 13. [23] Show that each of the following conditions implies the subsequent ones:
(1) $\mathbf{CC}(\mathbb{R})$.
(2) Lebesgue measure is σ–additive.
(3) $\mathbb{R}$ is not a countable union of countable sets.
[Hint: For the implication (1) $\Rightarrow$ (2) consult the proof of Proposition 7.14.]

E 14. [24] Show that:
(1) If X and Y are subsets of $\mathbb{R}$ with positive Lebesgue–measure each, then there exist $x \in X$ and $y \in Y$ with $(x - y) \in \mathbb{Q}$.
(2) Ugly functions are non–measurable.
[Hint: Use (1) to prove (2).]

E 15. Show that under each of the following conditions there exist non–Lebesgue–measurable subsets of $\mathbb{R}$:
(1) $\mathbf{CC}(\mathbb{R})$ holds and the Cantor cube $\mathbf{2}^{\mathbb{R}}$ is Weierstrass–compact.
(2) The Cantor cube $\mathbf{2}^{\mathbb{R}}$ is compact. [Hint: Exercises to Section 4.13, E 8 and E 9 above.]

E 16. [25] Show that:
(1) Under $\mathbf{AC}(\mathbb{R})$ there exist ugly functions with connected graphs.
(2) Graphs of ugly functins are never locally connected.
(3) Graphs of ugly isomorphisms are zerodimensional.
(4) Removal of any non–vertical straight line from the graph of an ugly function produces a totally disconnected remainder.

---

[22] [Rai84]

[23] In the Feferman–Levy Model A8 (M9 in [HoRu98]) $\mathbb{R}$ is a countable union of countable sets.

[24] [Sie20]

[25] [Jon42]

5.2 Disasters in Geometry: Paradoxical Decompositions

At first glance, the Banach–Tarski Decomposition seems preposterous. It blatantly contradicts our intuition about the conservation of mass or volume.

E. Schechter[26]

It certainly does seem to be folly to claim that a billiard ball can be chopped into pieces which can be put back together to form a life-size statue of Banach.

K. Stromberg[27]

I think there is still something **very** *disturbing about the Banach–Tarski paradox.*

S. Fefermann[28]

Intuition is an important guide for mathematicians. However, it is not always a safe one. There exist mathematical results that are counterintuitive. In many of these cases, a culprit can be isolated: the *Axiom of Choice*, i.e., many paradoxical results are demonstrable in **ZFC**, but not in **ZF**. The most stunning of these is the so called *Banach–Tarski Paradox* which establishes the existence of rather bizarre decompositions of the unit ball and of other heavenly bodies. The construction of these paradoxical decompositions is motivated by measure theoretic considerations. As we have seen in the previous section, the axiom of choice allows the construction of non–measurable subsets of the reals. The main obstacle — besides **AC** — turned out to be the requirement that the measure–function be σ–additive. What happens, if we relax this condition by requiring just additivity, but add the natural requirement that congruent (i.e., isometric) sets have the same measure? The paradoxical decompositions, unearthed by Hausdorff[29] and by Banach and Tarski[30] by means of the Axiom of Choice, demonstrate that in $\mathbb{R}^3$, the 3–dimensional space, even such functions, describing the *volume* of bounded bodies, do not exist. For $n = 1$ and $n = 2$ such measures do exist, as shown by Banach[31] — however (as we will point out later) they have some rather bizarre properties, as shown by von Neumann[32].

---

26 [Sch97, p. 142]
27 [Str79]
28 [Fef2000]
29 [Hau14]
30 [BaTa24]
31 [Bana23]
32 [vNeu29]

To prove these results is beyond the scope of this monograph. However, we will outline the main ideas, referring the interested reader for the more technical details to the elegant and *"strictly elementary account"* given in [Str79] or to the comprehensive book [Wag86], *"where this striking theorem* [the Banach–Tarski Paradox] *and many related results in geometry and measure theory, and the underlying tools of group theory, are presented with care and enthusiasm."*[33]

To describe the measure–theoretic results, concerning the spaces $\mathbb{R}^n$, we will use (in this section) the following terminology:

Definition 5.12. *For* $n \in \mathbb{N}^+$, *an* n–dimensional measure *is a function* $\mu_n \colon \mathcal{P}_b\mathbb{R}^n \to \mathbb{R}^+$, *defined on the set* $\mathcal{P}_b\mathbb{R}^n$ *of bounded subsets of* $\mathbb{R}^n$, *satisfying the following conditions:*

- *(M1)* μ_n *is* additive, *i.e.,* $\mu(A \cup B) = \mu(A) + \mu(B)$ *for disjoint elements* A *and* B *of* $\mathcal{P}_b\mathbb{R}^n$.
- *(M2)* μ *is* invariant, *i.e.,* $A \approx B$ *implies* $\mu_n(A) = \mu_n(B)$, *where* A *and* B *are called* congruent, *shortly* $A \approx B$, *iff there exists an isometry*[34] $f \colon \mathbb{R}^n \to \mathbb{R}^n$ *with* $f[A] = B$.
- *(M3)* μ_n *is* normed, *i.e.,* $\mu_n([0,1]^n) = 1$.

NOTE: In this section we work — unless stated otherwise — in **ZFC**

Hausdorff's Paradoxical Decomposition of the Sphere

Hausdorff[35] was the first to show that 3–dimensional measures (and hence n–dimensional measures for any $n \geq 3$; see Exercise E 1) do not exist. He obtained this result by exhibiting a paradoxical decomposition of the sphere:

5.13. Hausdorff's Decomposition Theorem for the Unit Sphere[36]
There exists a partition $\{A, B, C, D\}$ of the unit sphere
$S^2 = \{(x, y, z) \in \mathbb{R}^3 \mid x^2 + y^2 + z^2 = 1\}$ such that:

1. $A \approx B \approx C$.
2. $A \approx (B \cup C)$.
3. D is countable.

Later we will indicate the idea of the proof of the above Theorem. First, however, some consequences.

---

[33] [Wag86]. From the Foreword by Jan Mycielski.

[34] In fact, one could restrict attention to orientation–preserving isometries without changing any of the results in this section.

[35] [Hau14]

[36] [Hau14]

Corollary 5.14. [37] *There is no function $\mu\colon \mathcal{P}_b S^2 \to \mathbb{R}^+$ which satisfies (M1), (M2) and (M3') $\mu(S^2) > 0$.*

Proof. Assume that a function μ, satisfying (M1), (M2), and (M3'), exists. The countability of D implies that there exists an isometry f of the sphere S^2 (in fact a rotation) such that D and $f[D]$ are disjoint. Hence $D_1 = (D \cup f[D])$ is a countable subset of S^2 with $\mu(D_1) = 2 \cdot \mu(D)$. By repeating this process one obtains, for each $n \in \mathbb{N}^+$, a countable subset D_n of S^2 with $\mu(D_n) = 2^n \cdot \mu(D)$. Since $\mu(D_n) \leq \mu(S^2)$ for each n, this implies $\mu(D) = 0$. Consequently: $\mu(S^2) = \mu(A) + \mu(B) + \mu(C) = 3 \cdot \mu(A)$ and $\mu(S^2) = \mu(B \cup C) + \mu(B) + \mu(C) = 4 \cdot \mu(A)$. Therefore $\mu A = 0$, hence $\mu(S) = 0$, a contradiction.

Corollary 5.15. [38] *There is no 3–dimensional measure.*

Proof. If μ_3 would be a 3–dimensional measure, then the function $\mu\colon \mathcal{P}_b S^2 \to \mathbb{R}^+$, defined by

$$\mu(A) = \mu_3\big(\{(\lambda x, \lambda y, \lambda z) \mid (x, y, z) \in A \text{ and } 0 < \lambda \leq 1\}\big)$$

would satisfy (M1), (M2), and (M3'), contradicting Corollary 5.14.

Looking back at Hausdorff's Theorem 5.13 one observes that the existence of the countable set D somewhat reduces its elegance. Is there a smoother result? Sierpiński[39] showed that there are partitions
$\mathcal{P}_1 = \{A_1, \ldots, A_6, B_1, \ldots, B_4\}$,
$\mathcal{P}_2 = \{C_1, \ldots, C_6\}$ and
$\mathcal{P}_3 = \{D_1, \ldots, D_4\}$
of the sphere such that

1. $A_i \approx C_i$ for $i = 1, \ldots, 6$,
2. $B_i \approx D_i$ for $i = 1, \ldots, 4$.

Though here the countable set is avoided, the number of pieces is unnecessary large as shown by Robinson:

5.16. Robinson's Decomposition Theorem for the Unit Sphere[40] There exists a partition $\{A_1,\ A_2,\ B_1,\ B_2\}$ of the unit sphere into connected and locally connected pieces such that $A_1 \approx A_2 \approx A_1 \cup A_2$ and $B_1 \approx B_2 \approx B_1 \cup B_2$.

This is, in a way, the best (or worst?) possible result. As Robinson himself formulates[41]:

---

[37] [Hau14]
[38] [Hau14]
[39] [Sie48]. See also [BrCe75] and [Str79].
[40] [Rob47], [DekGr56].
[41] [Rob47]

"Thus we may cut S^2 into four pieces, and reassemble them in pairs to form two copies of S^2. We cannot use fewer than four pieces, since we cannot form a copy of S^2 out of a single piece which is not all of S^2. Thus for the surface problem, the minimum number of pieces in which to cut S^2 is four."

Bad Groups

After Banach and Tarski[42] had improved Hausdorff's construction to obtain simpler and more striking decompositions of 3–dimensional bodies (see below), von Neumann[43] showed that — besides **AC** — the structure of the group of isometries of $\mathbb{R}^3$ is responsible for the possibility of such paradoxical decompositions and thus the non-existence of 3–dimensional measures. This group contains a free group on two generators — and this fact causes all the trouble. In von Neumann's own words[44]:

Der Euklidische Raum scheint danach beim Erreichen der Dimensionszahl 3 jäh seinen Charakter zu ändern: für $n < 3$ läßt er einen allgemeinen Maßbegriff noch zu, für $n \geq 3$ nicht mehr!
Daß dem nicht so ist, daß vielmehr der innere Grund dieses sonderbaren Phänomens eine gewisse gruppentheoretische Eigenheit der n–dimensionalen Drehgruppe ist, dies zu zeigen, ist der Hauptzweck der vorliegenden Arbeit.

...

Der plötzliche Charakterwechsel des Euklidischen Raumes beim Erreichen und Überschreiten der Dimensionszahl 3 liegt einfach daran, daß die — bisher allein berücksichtigte — Gruppe O_n der längentreuen Abbildungen für $n = 1, 2$ "auflösbar" ist, für $n = 3, 4, \ldots$ hingegen eine freie Untergruppe mit zwei Erzeugenden σ, τ hat."

---

42 [BaTa24]

43 [vNeu29]

44 *Translation: "Apparently Euclidean space changes its character abruptly when reaching dimension 3: for $n < 3$ it allows a general concept of measure, for $n \geq 3$ this is no longer the case!*

To show that this is not so, that rather the deeper reason for this strange phenomenon is a specific group theoretic peculiarity of the n–dimensional isometry group, is the main purpose of the present article.

...

The abrupt change of character of Euclidean space when reaching and passing dimension 3 is simply caused by the fact that the group O_n of isometries — the only one that has been considered so far — is "solvable" for $n = 1, 2$, but contains a free group with two generators σ, τ for $n = 3, 4, \ldots$."

Definition 5.17. *1. The* free group F_2 on two generators a *and* b *is the set of all words* $x_1x_2\dots x_n$ *with letters* a, b, a^{-1}, *and* b^{-1} *such that* a *and* a^{-1} *are never adjacent and neither are* b *and* b^{-1}; *supplied with the following multiplication: If* $w = x_1 \dots x_n$ *and* $v = y_1 \dots y_m$ *are elements of* F_2, *then* $w \cdot v$ *is obtained in several steps:*

Step 1: Concatenate w *and* v *to obtain* $x_1 \dots x_n y_1 \dots y_m$.

Step 2: Remove x_n *and* y_1 *provided that* $\{x_n, y_1\} = \{a, a^{-1}\}$ *or* $\{x_n, y_1\} = \{b, b^{-1}\}$.

Step 3: Repeat step 2 as often as necessary until an element of F_2 *is obtained.*

The empty word, *i.e., the word with no letters, denoted sometimes by* Λ, *is the* neutral element *of* F_2.

2. $x \cdot Y = \{xy \mid y \in Y\}$ *for* $x \in F_2$ *and* $Y \subseteq F_2$.

3. Subsets X *and* Y *of* F_2 *are called* congruent, *in symbols* $X \approx Y$, *provided that there exists some* $z \in F_2$ *with* $Y = z \cdot X$.

Theorem 5.18. [45] *There exists a partition* $\{A, B, C, D\}$ *of the free group* F_2 *such that:*

1. $A \approx (A \cup C \cup D)$.
2. $C \approx (A \cup B \cup C)$.

Proof. To demonstrate simultaneously the main idea of the proof and the technical difficulty that has to be overcome, we present first an argument that almost works, next a complete proof.

Attempt: Define
$A = \{x_1 \dots x_n \in F_2 \mid x_1 = a\}$,
$B = \{x_1 \dots x_n \in F_2 \mid x_1 = a^{-1}\}$,
$C = \{x_1 \dots x_n \in F_2 \mid x_1 = b\}$,
$D = \{x_1 \dots x_n \in F_2 \mid x_1 = b^{-1}\}$.
Then $\{A, B, C, D\}$ is almost a partition of F_2. Just the empty word Λ is missing. Moreover:
$A \approx a^{-1} \cdot A = A \cup C \cup D \cup \{\Lambda\}$,
$C \approx b^{-1} \cdot C = A \cup B \cup C \cup \{\Lambda\}$.
So, the empty word muddles things up and causes a correct proof to be slightly less symmetric, hence less elegant:

Proof. Define A and B as above, but redefine C and D as follows:
$C = \{x_1 \dots x_n \in F_2 \mid x_1 = b\} \cup \{b^{-n} \mid n \in \mathbb{N}\}$[46],
$D = F_2 \backslash (A \cup B \cup C) = \{x_1 \dots x_n \in F_2 \mid x_1 = b^{-1}\} \backslash \{b^{-n} \mid n \in \mathbb{N}\}$.
Then $\{A, B, C, D\}$ is a partition of F_2, and:
$A \approx a^{-1} \cdot A = A \cup C \cup D$,
$C \approx b^{-1} \cdot C = A \cup B \cup C$.

---

45 [vNeu29]

46 Here $b^0 = \Lambda$.

Bad groups G that act *fixpoint-free* (i.e., only the neutral element has fixpoints) on a set X lead to paradoxical decompositions of X — where $A \approx B$ for subsets of X iff there exists $g \in G$ with $g[A] = B$:

Theorem 5.19. [47] *If F_2 acts fixpoint-free on X, then there exists a partition $\{A, B, C, D\}$ of X with*

1. $A \approx (A \cup C \cup D)$.
2. $C \approx (A \cup B \cup C)$.

Proof. Let $\{A, B, C, D\}$ be a partition of F_2 with $A \approx (A \cup C \cup D)$ and $(C \approx A \cup B \cup C)$. For each $x \in X$ let $\text{orb}(x) = \{g(x) \mid g \in F_2\}$ be the orbit of x. Then $\{\text{orb}(x) \mid x \in X\}$ is a partition of X. By **AC** there exists a subset S of X that contains exactly one element from each orbit. Define

$$A^* = \{g(x) \mid g \in A \text{ and } x \in S\}$$

and analogously B^*, C^*, and D^*. Since F_2 acts fixpoint-free on X, the set $\{A^*, B^*, C^*, D^*\}$ is a partition of X. Obviously:

$$A^* \approx (A^* \cup C^* \cup D^*) \text{ and } C^* \approx (A^* \cup B^* \cup C^*\}.$$

Paradoxical Decompositions of the Unit Ball

The above observations lead naturally to the following considerations: The group of all isometries of $\mathbb{R}^3$ contains a subgroup that is isomorphic to F_2 and acts on the unit ball $B_3 = \{(x, y, z) \in \mathbb{R}^3 \mid x^2 + y^2 + z^2 \leq 1\}$[48].

If this acting would be fixpoint-free, then the above results would immediately lead to a paradoxical decomposition of B_3 into 4 pieces. The fact, however, that rotations do have fixpoints — fortunately, not too many — causes complications. In fact, a paradoxical decomposition of B_3 in 4 pieces is impossible[49]. However, Banach and Tarski have been able to demonstrate the following:

---

47 [vNeu29]

48 Hausdorff, in his proof of the Decomposition Theorem 5.13, did not use a free subgroup of the isometry group of S^2, but rather one that is almost free, by showing that there exist rotations α and β of S^2 such that $\alpha^2 = \text{id} = \beta^3$ are the only relations in the group G generated by $\{\alpha, \beta\}$. Each element g of G has precisely two fixpoints. The union of these fixpoint-sets is Haudorff's countable set D. Then Hausdorff constructs judiciously a partition $\{A, B, C\}$ of G such that $\beta A = B$, $\beta^2 A = C$, and $\alpha A = B \cup C$. If X is a set obtained by selecting exactly one element from the orbit of each point $x \in (S^2 \setminus D)$, then the partition $\{A \cdot X, B \cdot X, C \cdot X\}$ of $S^2 \setminus D$ has the required properties.

49 [Rob47]

Theorem 5.20. [50] *There exist partitions*
$P_1 = \{A_1, \ldots, A_n, B_1, \ldots, B_m\}$,
$P_2 = \{C_1, \ldots, C_n\}$, *and*
$P_3 = \{D_1, \ldots, D_m\}$ *of the unit ball such that:*

1. $A_i \approx C_i$ *for* $i = 1, \ldots, n$.
2. $B_i \approx D_i$ *for* $i = 1, \ldots, m$.

How many pieces $n + m$ are needed to "double" a ball?

- Stromberg[51] showed that $40 = 24 + 16$ suffice,
- Bruckner and Ceder[52] used $30 = 18 + 12$,
- von Neumann[53] used $9 = 5 + 4$,
- Sierpiński[54] used $8 = 5 + 3 = 6 + 2$, and
- Robinson[55] supplied the ultimate answer $5 = 3 + 2$:

5.21. Robinson's Decomposition Theorem for the Unit Ball[56]
There exist partitions
$P_1 = \{A_1, A_2, A_3, B_1, B_2\}$,
$P_2 = \{C_1, C_2, C_3\}$ and
$P_3 = \{D_1, D_2\}$
of the unit ball into connected and locally connected pieces such that

1. $A_i \approx C_i$ for $i = 1, 2, 3$.
2. $B_i \approx D_i$ for $i = 1, 2$.

Moreover, Robinson showed that 4 pieces do not suffice.

The Banach–Tarski Paradox

Since, by the last two theorems, any ball in 3–dimensional space can be "doubled", it follows easily that for any bounded subset A of $\mathbb{R}^3$ and any ball B in $\mathbb{R}^3$ there exist a partition $\{A_1, \ldots, A_n\}$ of A and a partition $\{B_1, \ldots, B_n\}$ of some subset of B such that $A_i \approx B_i$ for each $i = 1, \ldots, n$ (see Exercise E 2). Thus for subsets A and B of $\mathbb{R}^3$ that are bounded and contain some ball each, it is possible to decompose each into a finite number of pieces and to "reassemble" these pieces to form a subset of the other set.

The Banach–Tarski Paradox says even more. To state it properly, a definition first:

---

50 [BaTa24]
51 [Str79]
52 [BrCe75]. See also [Sie48].
53 [vNeu29]
54 [Sie45]
55 [Rob47]
56 [Rob47], [DekGr56].

Definition 5.22. [57] *Subsets A and B of $\mathbb{R}^3$ are called* equidecomposable *in symbols, $A \sim_e B$, iff there exist partitions $\{A_1, \ldots, A_n\}$ of A and $\{B_1, \ldots, B_n\}$ of B with $A_i \approx B_i$ for $i = 1, \ldots, n$.*

5.23. The Banach–Tarski Paradox[58] Any two bounded subsets A and B of $\mathbb{R}^3$, that contain some ball each, are equidecomposable.

Proof. By the above remarks, each of the sets A and B is equidecomposable to some subset of the other. Thus the result follows immediately from the next theorem.

Theorem 5.24. [59] *If subsets A and B of $\mathbb{R}^3$ are each equidecomposable to some subset of the other, then A and B are equidecomposable.*

Proof. Let $\{A_1, \ldots, A_n\}$ be a partition of A and let $\{B_1, \ldots, B_n\}$ be a partition of a subset B' of B with $A_i \approx B_i$ for each i. Then for each $i = 1, \ldots, n$ there exists an isometry $f_i\colon A_i \to B_i$. Thus the map $f\colon A \to B'$, defined by $f(a) = f_i(a)$ for $a \in A_i$, is a bijection satisfying the condition:

(A) $C \sim_e f[C]$ for each subset C of A.

Likewise there exists a bijection $g\colon B \to A'$ from B to some subset A' of A, satisfying the condition

(B) $D \sim_e g[D]$ for each subset D of B.

Define, via recursion, a sequence (C_n) of subsets C_n of A by

$$\begin{cases} C_0 &= A \setminus A' \\ C_{n+1} &= g\big[f[C_n]\big]. \end{cases}$$

Consider $C = \bigcup_{n\in\mathbb{N}} C_n$. Then a simple computation shows that $A\setminus C = g\big[B\setminus f[C]\big]$. Thus condition (B) implies $(A\setminus C) \sim_e (B\setminus f[C])$. Since (A) implies $C \sim_e f[C]$, it follows that $A = (A \setminus C) \cup C \sim_e (B \setminus f[C]) \cup f[C] = B$.

AC as the Culprit

As we have seen above, in **ZFC** one can prove the existence of rather counterintuitive and undesirable decompositions of balls and other bodies in 3–dimensional space. Is **AC** really the culprit or are similar decompositions also constructible in **ZF**? Since Lebesgue–measure in $\mathbb{R}^3$ is additive, invariant and normed, the above paradoxes imply that there are bounded subsets of $\mathbb{R}^3$ that are not Lebesgue–measurable. So the fact that there exist models[60] of **ZF** in which all bounded subsets of $\mathbb{R}^3$ are Lebesgue–measurable, shows that here again **AC** is the villain.

---

57 [BaTa24]

58 [BaTa24]

59 [Bana23]. Observe that this result holds in **ZF**.

60 E.g., Shelah's Second Model A2 (M38 in [HoRu98]).

The following diagram shows how the existence of paradoxical decompositions is related to other axioms, including **HBT**, the *Hahn–Banach Theorem*, considered to be *"The Crown Jewel of Functional Analysis"*:

Diagram 5.25. [61] In **ZF** the following implications hold:

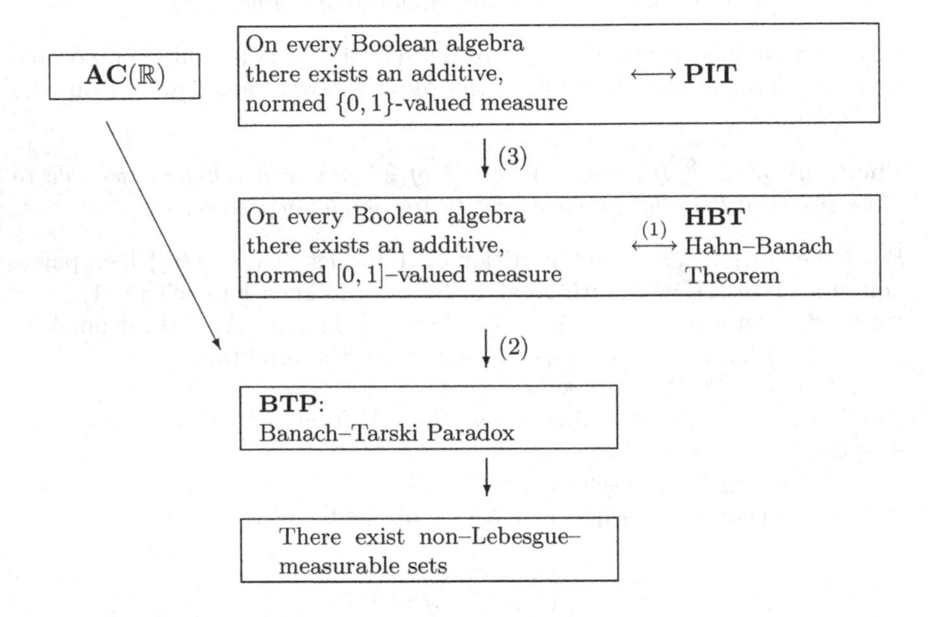

All implications with the possible exception of the penultimate one (2) are proper.

Bizarre Decomposition Paradoxes in Dimensions 1 and 2

Although — as Banach[62] has shown — there exist measures on $\mathbb{R}^2$ (invariant under isometries) — even different ones, some agreeing on all Lebesgue–measurable sets with the Lebesgue measure, others failing to do that; there do exist paradoxical decompositions — as von Neumann[63] has shown — provided we enlarge the isometry group to the group A_2 of all area–preserving affine maps (i.e., those with determinant 1):

5.26. Von Neumann–Lemma A_2 contains a subgroup that is free on two generators.

[61] For (1) see [Lux69]. For (2) see [Paw91] and [FoWe91]. For (3) see [LoRy51]. For detailed accounts of the Hahn–Banach Theorem see [Bus93] and [NaBe97].

[62] [Bana23]

[63] [vNeu29]

5.27. Decomposition Paradox for the Plane Any two bounded planar sets A and B with non–empty interiors are A_2–equidecomposable, i.e., there exist partitions $\{A_1, \ldots, A_n\}$ of A and $\{B_1, \ldots, B_n\}$ of B such that for each $i \in \{1, \ldots, n\}$ there exists an affine transformation $T_i \colon \mathbb{R}^2 \to \mathbb{R}^2$ with determinant 1 such that $t_i[A_i] = B_i$.

Even though in $\mathbb{R}^1$, i.e., on the real line, the only continuous transformations that preserve distances are the translations and the reflections, which do not give rise to paradoxical decompositions, von Neumann has been able to unearth the following linear decomposition paradox:

5.28. Von Neumann's Decomposition Paradox for the Real Line[64] For any two bounded linear sets A and B with non–empty interiors there exist partitions $\{A_1, \ldots, A_n\}$ of A and $\{B_1, \ldots, B_n\}$ of B such that for each $i \in \{1, \ldots, n\}$ there exists a bijection $f_i \colon A_i \to B_i$ such that for any two points x and y of A_i we have $|x - y| < |f(x) - f(y)|$.

Corollary 5.29. [65] *For any 1–dimensional measure μ there exist bounded linear sets A and B and a bijection $f \colon A \to B$ that increases the distances of any two points of A, but such that $\mu(B) < \mu(A)$.*

5.30. Sierpiński's Decomposition Paradox for Disks[66]
For any pair (r, s) of positive reals there exist partitions $\{A_1, \ldots, A_n\}$ of $\{(x, y) \in \mathbb{R}^2 \mid x^2 + y^2 \le r^2\}$ and $\{B_1, \ldots, B_n\}$ of $\{(x, y) \in \mathbb{R}^1 \mid x^2 + y^2 \le s^2\}$ such that for each $i \in \{1, \ldots, n\}$ there exists a bijection $f_i \colon A_i \to B_i$ such that for any two points a and b of A_i we have $d(a, b) < d\big(f(a), f(b)\big)$.

Exercises to Section 5.2:

E 1. Show that, if $n \le m$, then the existence of an m–dimensional measure implies the existence of an n–dimensional measure in **ZF**.

E 2. Let A be a bounded subset of $\mathbb{R}^3$ and B be a subset of $\mathbb{R}^3$ that contains some ball. Show that A is equidecomposable with some subset of B.

E 3. Show that Banach's Theorem 5.24 holds in **ZF**.

E 4. [67] Some Non–paradoxical but peculiar decomposition in **ZF**.
Let $\mathbb{P}$ (resp. $\mathbb{T}$) be the set of all irrational (resp. transcendental) real numbers. Show that:
(1) There exist partitions
$\{R_1, R_2\}$ of $\mathbb{R}$ and $\{P_1, P_2\}$ of $\mathbb{P}$ with $R_1 \approx P_1$ and $R_2 \approx P_2$.

---

[64] [vNeu29]
[65] [vNeu29]
[66] [Sie48]
[67] [Sie48]

(2) There exist partitions
$\{P_1, P_2\}$ of $\mathbb{P}$ and $\{T_1, T_2\}$ of $\mathbb{T}$ with $P_1 \approx T_1$ and $P_2 \approx T_2$.

(3) There do not exist partitions
$\{Q_1, Q_2, \ldots, Q_n\}$ of $\mathbb{Q}$ and $\{A_1, A_2, \ldots, A_n\}$ of $\mathbb{R} \setminus \mathbb{T}$ such that $Q_i \approx A_i$ for $i \in \{1, 2, \ldots, n\}$.

[Hint: For $a \in \mathbb{R}$ let $t_a \colon \mathbb{R} \to \mathbb{R}$ be defined by $t_a(x) = a + x$.
Re (1): Observe that for $a \in \mathbb{R}$ and $X = \bigcup_{n \in \mathbb{N}} t_a^n[\mathbb{Q}]$ we get $t_a[X] = X \setminus \mathbb{Q}$.
Re (2): Observe that for $a \in \mathbb{T}$ and $X = \bigcup_{n \in \mathbb{N}} t_a^n[\mathbb{R} \setminus (\mathbb{T} \cup \mathbb{Q})]$ we get
$t_a[X] = X \setminus \big(\mathbb{R} \setminus (\mathbb{T} \cup \mathbb{Q})\big)$.
Re (3): Observe that $|x - y| \in \mathbb{Q}$ for x and y in $\mathbb{Q}$, but $|n \cdot \sqrt{2} - m \cdot \sqrt{2}\,| \notin \mathbb{Q}$ for n and m different natural numbers.]

6

Disasters either way

There are two kinds of truth. To the one kind belong statements so simple and clear that the opposite assertions obviously could not be defended. The other kind, the so-called "deep truths", are statements in which the opposite also contains deep truth.

Niels Bohr[1]

6.1 Disasters in Game Theory

The axiom of choice may well be regarded as such a "deep truth". Its Janus faced nature is dramatically revealed by the theory of games. On one hand **AC** guarantees the existence of winning strategies for certain deterministic 2–person games with complete information; on the other hand **AC** allows the "construction" of similar deterministic 2–person games with complete information that lack winning strategies.

Let us start by introducing the relevant concepts via the description of the following simple games:

Definition 6.1. *The 2–player* game

$$G = G\big(n, (X_1, \ldots, X_n),\ (Y_1, \ldots Y_n),\ A\big)$$

where

- *n is a positive integer, the number of* moves *of each of the two players.*
- *$(X_1, \ldots, X_n)$ (resp. $(Y_1, \ldots, Y_n)$) is an n–tuple of non-empty sets X_i (resp. Y_i) whose elements are the possible i–th* moves *of the first (resp. second) player.*
- *A is a subset of $\prod_{i=1}^{n} (X_i \times Y_i)$, called the* winning set *for the second player.*

[1] [Boh49]. See also [Myc66].

The game G is played *as follows: The players choose successively elements*

$$x_1 \in X_1,\ y_1 \in Y_1,\ x_2 \in X_2, \dots, y_n \in Y_n$$

at each step knowing all the previous steps. The $2n$–tuple $(x_1, y_1, x_2, \dots, y_n)$ is called the outcome *of the game.*

The second player wins if the outcome belongs to A; otherwise the first player wins.

A strategy *for the first player is an n–tuple $\sigma = (\sigma_1, \dots, \sigma_n)$ of functions*[2]

$$\sigma_i \colon \prod_{j<i} (X_j \times Y_j) \to X_i.$$

A strategy σ is called a winning strategy for the first player *provided that for any n–tuple $(y_1, \dots, y_n) \in \prod_{i=1}^{n} Y_i$ the $2n$–tuple $(x_1, y_1, x_2, \dots, y_n)$, defined by $x_1 = \sigma(\emptyset)$ and $x_{i+1} = \sigma(x_1, y_1, x_2, \dots, y_i)$ for all $i = 1, \dots, n-1$, does not belong to A, i.e., that the first player wins the game provided that he uses the strategy σ, no matter what the second player does.*

Similarly strategies and winning strategies are defined for the second player.

The game G is called determinate *provided that one of the players has a winning strategy.*

For finite games of the above form[3] all is well:

Theorem 6.2. *The game $G = G\big(n, (X_1, \dots, X_n),\ (Y_1, \dots, Y_n),\ A\big)$ is determinate if all the X_i's and Y_i's are finite.*

Proof. We proceed by induction.

Step 1: $n = 1$.

Two cases are possible:

Case 1: There exists $x_1 \in X_1$ with $(\{x_1\} \times Y_1) \cap A = \emptyset$. Then the first player has a winning strategy by choosing such an x_1.

Case 2: For each $x_1 \in X_1$ there exists some $y_1 \in Y_1$ with $(x_1, y_1) \in A$. Since X_1 is finite, there exists a function $\sigma \colon X_1 \to Y_1$ such that $(x, \sigma(x)) \in A$ for each $x \in X$. Such σ provides a winning strategy for the second player.

Step 2: Assume that each game $G = G(n, \dots)$ with $n \leq k$ is determinate. Consider a game $G = G(k+1, (X_1, \dots, X_{k+1}),\ (Y_1, \dots, Y_{k+1}),\ A)$. Then for each pair $(x, y) \in (X_1 \times Y_1)$ we get a new game $G(x, y) = G\big(k, (X_2, \dots, X_{k+1}),\ (Y_2, \dots, Y_{k+1}),\ A(x, y)\big)$, where

---

[2] $\{\emptyset\}$ is the empty product.

[3] Many familiar deterministic 2–person games with complete information can be represented in the above form. If ties are possible, like in chess, minor adjustments are needed. See Exercise E 3.

$$A(x,y) = \Big\{\big((x_2,y_2),\ldots,(x_{k+1},y_{k+1})\big) \in \prod_{i=2}^{k+1}(X_i \times Y_i) \mid \big((x,y),\ (x_2,y_2),\ldots,(x_{k+1},y_{k+1})\big) \in A\Big\}$$

Then two cases are possible:

Case 1: There exists $x_1 \in X_1$ such that for each $y \in Y_1$ the first player has a winning strategy $\sigma(y)$ for the game $G(x_1, y)$.
Since Y_1 is finite, this implies that the first player has a winning strategy for the original game G.

Case 2: For each $x \in X_1$ there exists some $y \in Y_1$ such that the second player has a winning strategy $\sigma(x,y)$ for the game $G(x,y)$.
Since $X_1 \times Y_1$ is finite, this implies that the second player has a winning strategy of the original game G.

Problems arise when we pass from finite to infinite games. There are two natural ways to do this. We may allow, for each player, ω moves instead of a finite number n only or we may allow the players an infinite number of options for some of their moves, i.e., we may allow the X_i's and Y_i's to be infinite. In both cases the games may loose their determinateness:

Disaster 6.3. Infinite games of the form

$$G\big(\omega, (X_n)_{n\in\omega},\ (Y_n)_{n\in\omega}, A\big) \text{ resp. } G\big(n, (X_i)_{i\le n}, (Y_i)_{i\le n}, A\big)$$

may fail to be determinate.

The reasons for the disaster concerning the two types of infinite games described above are decidedly complementary to each other.

Theorem 6.4. *Equivalent are:*

1. *Each game of the form* $G\big(n, (X_1,\ldots,X_n),\ (Y_1,\ldots,Y_n),\ A\big)$ *is determinate.*
2. *Each game of the form* $G\big(1, (X_1),\ (Y_1),\ A\big)$ *is determinate.*
3. **AC**.

Proof. **(1)** $\Rightarrow$ **(2)** Obvious.

(2) $\Rightarrow$ **(3)** Let $(X_i)_{i\in I}$ be a family of non–empty sets. Consider $X = \bigcup_{i\in I} X_i$ and $A = \{(i,x) \mid i \in I \text{ and } x \in X_i\}$. Then the game $G = G\big(1,(I),(X),A\big)$ is determinate, by (2). Since for every $i \in I$ there exists $x \in X$ with $(i,x) \in A$, the first player can have no winning strategy. Thus the second player must have a winning strategy, i.e., a function $\sigma\colon I \to X$ such that $\big(i,\sigma(i)\big) \in A$ for each $i \in I$, — in other words: $\sigma \in \prod_{i\in I} X_i$. Thus **AC** holds.

(3) $\Rightarrow$ **(1)** This implication is verified as in the proof of Theorem 6.2, since in the presence of **AC** the finiteness–assumptions are not needed.

Theorem 6.5. [4] *In* **ZFC** *there exists a subset* A *of* $(\{0,1\}^2)^{\mathbb{N}}$ *such that the game*
$G_A = (\omega, \{0,1\}^{\mathbb{N}}, \{0,1\}^{\mathbb{N}}, A)$ *is not determinate.*

Proof. [5] Independently of A, each player has at each step precisely 2 options to play, namely 0 and 1, thus altogether $2^{\aleph_0}$ options. Since $(2^{\aleph_0})^{\aleph_0} = 2^{\aleph_0}$, a simple computation shows that each player has — independently of A — precisely $2^{\aleph_0}$ possible strategies to play the game. By **AC**, $2^{\aleph_0} = \aleph_\gamma$ for some ordinal γ. Thus the possible strategies for the first (resp. second) player can be arranged in the form $(\sigma_\alpha)_{\alpha<\aleph_\gamma}$ (resp. $(\tau_\alpha)_{\alpha<\aleph_\gamma}$). By transfinite recursion we will construct, for $\alpha < \aleph_\gamma$, subsets A_α and B_α of $(\{0,1\}^2)^{\mathbb{N}}$ such that

1. $\alpha < \beta \Rightarrow (A_\alpha \subseteq A_\beta$ and $B_\alpha \subseteq B_\beta)$,
2. $|A_\alpha| \leq |\alpha|$ and $|B_\alpha| \leq \alpha$,
3. $A_\alpha \cap B_\alpha = \emptyset$,

such that the game G_A with $A = \bigcup_{\alpha<\aleph_\gamma} A_\alpha$ is not determinate.

Let us assume that the A_α's and B_α's are constructed for $\alpha < \beta$ according to the above restrictions:

Case 1: $\beta = 0$
Choose $A_0 = B_0 = \emptyset$

Case 2: β is a limit ordinal.
Choose $A_\beta = \bigcup_{\alpha<\beta} A_\alpha$ and $B_\beta = \bigcup_{\alpha<\beta} B_\alpha$.

Case 3: $\beta = \alpha + 1$ for some α.
If the first player plays according to the strategy σ_α and the second player plays $y = (y_1, y_2, \ldots)$ for some $y \in \{0,1\}^{\mathbb{N}}$ the outcome will be of the form $0(\sigma_\alpha, y) = (x_1, y_1, x_2, y_2, \ldots)$. Since $|\{0,1\}^{\mathbb{N}}| = 2^{\aleph_0}$ and $|B_\alpha| < 2^{\aleph_0}$, there exist some $y(\sigma_\alpha)$ in $\{0,1\}^{\mathbb{N}}$ such that $0(\sigma_\alpha, y(\sigma_\alpha)) \notin B_\alpha$. Select such an element $y(\sigma_\alpha)$ and define $A_\beta = A_\alpha \cup \{0(\sigma_\alpha, y(\sigma_\alpha))\}$. This implies that σ_α is not a winning strategy of the first player for the game G_{A_β}, and thus also not for the original game G_A. Similarly there exists some $x(\tau_\alpha)$ in $\{0,1\}^{\mathbb{N}}$ such that, if the first player plays $x(\tau_A)$ and the second player plays according to the strategy τ_α, the outcome $0(x(\tau_\alpha), \tau_\alpha)$ will not belong to A_β. Define $B_\beta = B_\alpha \cup \{0(x(\tau_\alpha), \tau_\alpha)\}$. This implies that τ_α is not a winning strategy of the second player for the game $G_{(\{0,1\}^2)^{\mathbb{N}} \setminus B_\beta}$, and thus also not for the game G_A. Consequently the game G_A is not determinate.

The *Axiom of Determinateness*, **AD**, stating that the above game G_A is determinate for each A, will be investigated further in Section 7.2.

[4] [Myc64]
[5] [Jec73]

Exercises to section 6.1:

E 1. Show that if in a game of the form $G\big(1,(X_1),\ (Y_1),\ A\big)$ the first player has no winning strategy, then the second player can *always* win, even though he may not have a winning strategy.

E 2. Show the equivalence of:
(1) The game $G(1,(\mathbb{N}),(\mathbb{R}),A)$ is determinate for each set $A \subseteq (\mathbb{N} \times \mathbb{R})$.
(2) $\mathbf{CC}(\mathbb{R})$.

E 3. Consider the following modifications of the game $G\big(n,(X_1,\dots,X_n),\ (Y_1,\dots,Y_n),\ A\big)$: Replace A by a partition (A,B,C) of the set $\prod_{i=1}^{n}(X_i \times Y_i)$, and stipulate that
- the second player wins, if the outcome of a game belongs to A,
- the first player wins, if the outcome of a game belongs to B,
- there is a tie, if the outcome of the game belongs to C.

Show that for this game:
(1) The second player has a winning strategy iff he has a winning strategy for the game $G\big(n,(X_1,\dots,X_n),\ (Y_1,\dots,Y_n),\ A\big)$.
(2) The first player has a winning strategy iff he has a winning strategy for the game $G\big(n,(X_1,\dots,X_n),\ (Y_1,\dots,Y_n), A\cup C\big)$.
(3) Both players have strategies guaranteeing at least a tie iff the second player has a winning strategy for the game $G\big(n,(X_1,\dots,X_n),\ (Y_1,\dots,Y_n),\ A\cup C\big)$ and the first player has a winning strategy for the game $G\big(n,(X_1,\dots,X_n),\ (Y_1,\dots,Y_n),\ A\big)$.

E 4. Consider the constant sequence (0) with value 0 and the set $A = \{(x_n,y_n) \in (\{0,1\}^2)^{\mathbb{N}} \mid \forall_n\ y_n = x_n\}$. Show that:
(1) The second player has a winning strategy for the game $\big(\omega,\{0,1\}^{\mathbb{N}},\ \{0,1\}^{\mathbb{N}}\ ,A\big)$.
(2) The first player has a winning strategy for the game $\big(\omega,\{0,1\}^{\mathbb{N}},\ \{0,1\}^{\mathbb{N}},\ A\cup\{(0)\}\big)$.

E 5. Consider the following 2–person game H_A, where A is a subset of the interval $[0,2]$: Both players choose successively elements $x_0, y_0, x_1, y_1, x_2, y_2, \dots$ of $\{0,1\}$. The second player wins, if $\sum_{n=0}^{\infty}\big(\frac{x_n}{2^{2n}}+\frac{y_n}{2^{2n+1}}\big)$ belongs to A; otherwise the first player wins. Use Theorem 6.5 to show that in **ZFC** there exist subsets A of $[0,2]$ such that H_A is not determinate.

E 6. Discuss, whether a game of the form $G(\omega,(X_n)_{n\in\mathbb{N}},\ (Y_n)_{n\in\mathbb{N}}, A)$ "can be played" in **ZF** if $\prod_{n\in\mathbb{N}} X_n$ or $\prod_{n\in\mathbb{N}} Y_n$ happens to be empty.

7

Beauty without Choice

It seems that the well–known arguments against the axiom of choice have been exploited until today only in a negative sense.

J. Mycielski and H. Steinhaus[1]

The analogy with Geometry, ..., suggests the question: what shape will analysis and set theory assume by accepting a principle **contradicting** *the axiom of choice? Such a "non–Zermelian" theory in some sense corresponds to non–Euclidean geometry.*

A.A. Fraenkel, Y. Bar–Hillel and A. Levy[2]

7.1 Lindelöf = Compact

For me the proof of a theorem by means of Zermelo's axiom is valuable only as an indication that it is useless to waste time on an exact proof of the falsity of the theorem in question.

N. Lusin[3]

Aber hier, wie überhaupt, kommt es anders, als man glaubt.

Wilhelm Busch[4]

---

1 [MySt62]
2 [FrBaLe73, p. 85/86]
3 Cited after [Sie58, p. 95].
4 From: *Plisch und Plum.*

You have only to show that a thing is impossible and some mathematician will go and do it.

A saying[5]

In this section will be demonstrated that Lusin's verdict above is false, when reformulated as follows:

The proof of the falsity of some statement by means of the Axiom of Choice is valuable only as an indication that it is useless to waste time on an exact proof of the statement itself.

In fact, it will be shown that the following statements, each being false in **ZFC**, will become true theorems under the assumption that **AC** is badly false:

Disaster 7.1. The following statements are false in **ZFC**:

1. Products of Lindelöf T_2–spaces are Lindelöf.
2. Finite products of Lindelöf T_1–spaces are Lindelöf.
3. Lindelöf T_2–spaces are regular.
4. Totally disconnected Lindelöf T_2–spaces are zerodimensional.

Proof. See [Eng89] or the Theorems 7.4, 7.6, 7.7, 7.8 below.

The above failures of the Lindelöf property to behave nicely are particularly unfortunate in view of the fact that the Lindelöf property occupies a prominent place in **ZFC**–topology, in particular[6]

(a) All compact spaces (more generally: all σ–compact spaces[7]) and all second countable spaces in particular, all separable metrizable spaces (more generally: all separable, paracompact spaces) are Lindelöf.
(b) All regular Lindelöf spaces are paracompact and realcompact.
(c) Every locally compact, paracompact space is a sum of locally compact, Lindelöf T_2–spaces, and vice versa.
(d) For metrizable spaces: Lindelöf = separable.
(e) Continuous images, closed subspaces and countable sums of Lindelöf spaces are Lindelöf.

As the above observations indicate, the Lindelöf property behaves almost as compactness, one of the main differences being that compactness behaves much better than the Lindelöf property with respect to the formation of products.

Here now a big surprise:

5 Taken from [Saw82, p. 167].
6 See, e.g., [Eng89].
7 See Exercise E 1.

Theorem 7.2. [8] *Equivalent are:*

1. *For T_1-spaces: Lindelöf = compact.*
2. *For subspaces of $\mathbb{R}$: Lindelöf = compact.*
3. $\mathbf{CC}(\mathbb{R})$ *fails.*

Proof. (**1**) $\Rightarrow$ (**2**) Trivial.

(**2**) $\Rightarrow$ (**3**) If (2) holds, then $\mathbb{N}$ is not Lindelöf. Thus, by Theorem 3.8, $\mathbf{CC}(\mathbb{R})$ fails.

(**3**) $\Rightarrow$ (**1**) We need only show that failure of (1) implies $\mathbf{CC}(\mathbb{R})$. So let X be a non–compact, Lindelöf T_1-space. Let $\mathcal{C}$ be an open cover of X that has no finite subcover. Since X is Lindelöf we may assume $\mathcal{C}$ to be countable. By forming finite unions and deleting superfluous members we obtain an open cover $\mathcal{L} = \{B_n \mid n \in \mathbb{N}\}$ of X such that

- $B_n \subseteq B_m$ for $n \leq m$ and
- $C_n = (B_n \setminus \bigcup_{m<n} B_m) \neq \emptyset$ for each $n \in \mathbb{N}$.

Define, for each $n \in \mathbb{N}$ and each $x \in C_n$, the set

$$A(n, x) = B_n \setminus \{x\}$$

and consider the open cover

$$\mathfrak{A} = \{A(n, x) \mid n \in \mathbb{N} \text{ and } x \in C_n\}$$

of X. Then there exist unique maps $\alpha\colon \mathfrak{A} \to \mathbb{N}$ and $\beta\colon \mathfrak{A} \to X$ such that $A = A\big(\alpha(A), \beta(A)\big)$ for each $A \in \mathfrak{A}$.

Since X is Lindelöf, $\mathfrak{A}$ has a countable subcover $\{A_n \mid n \in \mathbb{N}\}$. The set $M = \{\alpha[A_n] \mid n \in \mathbb{N}\}$ is an unbounded, thus countable subset of $\mathbb{N}$. For each $m \in M$ define $x_m = \beta(A_{\min\{n \in \mathbb{N} \mid \alpha(A_n) = m\}})$. Then $x_m \in C_m$. The subspace Y of X with underlying set $\{x_m \mid m \in M\}$ is countable and discrete, since for each $m \in \mathbb{N}$

(a) the set $\{x_n \mid n \leq m\} = B_m \cap Y$ is open in Y,
(b) the set $\{x_n \mid n < m\}$ is closed in Y as a finite subset of a T_1-space,

and thus

(c) $\{x_m\}$ is open in Y.

Consequently Y is homeomorphic to the discrete space $\mathbb{N}$. Moreover, each element x of X is contained in some B_n, and thus has a neighborhood that meets Y in a finite set. Hence the T_1-property of X implies that Y is closed in X and thus Lindelöf (cf. Exercises to Section 3.2, E 1). Consequently $\mathbb{N}$ is Lindelöf, and Theorem 3.8 implies that $\mathbf{CC}(\mathbb{R})$ holds.

---

[8] [Her2002]

Corollary 7.3. [9] *Every* **ZF***–model satisfies exactly one of the following two alternatives:*

1. *Every subspace of* $\mathbb{R}$ *is Lindelöf.*
2. *For subspaces of* $\mathbb{R}$*: Lindelöf = compact = closed and bounded.*

Proof. If $\mathbf{CC}(\mathbb{R})$ holds then, by Theorems 4.54, condition (1) holds true. If $\mathbf{CC}(\mathbb{R})$ fails then, by Theorems 7.2 and 4.52, condition (2) holds true.

As pointed out earlier each of the two above cases can occur.

Theorem 7.4. [10] *Equivalent are:*

1. *Products of Lindelöf* T_2*–spaces are Lindelöf.*
2. **PIT** *holds and* $\mathbf{CC}(\mathbb{R})$ *fails.*

Proof. **(1)** $\Rightarrow$ **(2)** Consider the space $\mathbb{N}^{\mathbb{R}}$. Let $\mathcal{P}_2\mathbb{R}$ be the set of all subsets of $\mathbb{R}$ with exactly two elements. For $D = \{a, b\}$ in $\mathcal{P}_2\mathbb{R}$ define

$$C(D) = \{(n_x) \in \mathbb{N}^{\mathbb{R}} \mid n_a = n_b\}.$$

Since $\mathbb{R}$ is uncountable, the set $\mathcal{C} = \{C(D) \mid D \in \mathcal{P}_2\mathbb{R}\}$ is an open cover of $\mathbb{N}^{\mathbb{R}}$. However $\mathcal{C}$ has no countable subcover of $\mathbb{N}^{\mathbb{R}}$. To see this, consider a sequence (D_n) in $\mathcal{P}_2\mathbb{R}$. Then $D = \bigcup_{n\in\mathbb{N}} D_n$ is at most countable. Thus there exists an injective map $f \colon D \to \mathbb{N}$. Consequently the point (n_x) of $\mathbb{N}^{\mathbb{R}}$, defined by $n_x = \begin{cases} f(x), & \text{if } x \in D \\ 0, & \text{otherwise} \end{cases}$ does not belong to $\bigcup_{n\in\mathbb{N}} C(D_n)$. This fact implies that $\mathbb{N}^{\mathbb{R}}$ fails to be Lindelöf. Thus (1) implies that $\mathbb{N}$ fails to be Lindelöf. So, by Theorem 3.8, $\mathbf{CC}(\mathbb{R})$ fails. Consequently, by Theorem 7.2, the equality Lindelöf = compact holds for T_1–spaces and thus in particular for T_2–spaces. So, by (1) products of compact T_2–spaces are compact. Thus Theorem 4.70 implies that **PIT** holds.

(2) $\Rightarrow$ **(1)** If $\mathbf{CC}(\mathbb{R})$ fails, Theorem 7.2 implies, as above, that Lindelöf = compact for T_2–spaces. Thus **PIT** implies, via Theorem 4.70, that products of Lindelöf T_2–spaces are Lindelöf.

Remark 7.5. [11]

1. Observe that there exist models[12] of **ZF** in which **PIT** holds but $\mathbf{CC}(\mathbb{R})$ fails. Thus in these models the theorem

   Products of Lindelöf T_2–spaces are Lindelöf

   holds true. Since the Lindelöf–property is closed–hereditary (cf. Exercises to Section 3.2, E 1), in the above models the Lindelöf T_2–spaces form an

9 [Her2002]
10 [Her2002]
11 [Her2002]
12 Cohen's First Model A4 (M1 in [HoRu98]).

epireflective subcategory, i.e., — besides a *Tychonoff Theorem* — there is also a *Čech–Stone Theorem* for Lindelöf spaces. In particular, in these models, the familiar Čech–Stone compactification $\mathbb{N} \hookrightarrow \beta\mathbb{N}$ of $\mathbb{N}$ is the Lindelöf–T_2–reflection of $\mathbb{N}$ — somewhat surprising, perhaps.

2. Note further that there is no model of **ZF** in which products of Lindelöf T_1–spaces are always Lindelöf. This can be seen as follows: By Theorem 7.4, in such a model **CC**($\mathbb{R}$) must fail and products of compact T_1–spaces must be compact. Hence (see Exercises to Section 4.8, E 11) **AC** must hold. But if **AC** holds, then **CC**($\mathbb{R}$) cannot fail, a contradiction.

The situation is even worse for T_0–spaces. See Exercise E 2.

What about finite products?

Theorem 7.6. *Equivalent are:*

1. *Finite products of Lindelöf T_1–spaces are Lindelöf.*
2. **CC**($\mathbb{R}$) *fails.*

Proof. (**1**) ⇒ (**2**) Consider the *Sorgenfrey line* $S = (\mathbb{R}, \sigma)$, i.e., the topological space that has $\mathbb{R}$ as underlying set and the collection of all intervals of the form

$$[a, b) = \{x \in \mathbb{R} \mid a \leq x < b\}$$

as a base for the topology σ. Then σ is finer than the canonical topology τ on $\mathbb{R}$. In particular S is a T_1–space. Moreover, the space S^2 fails to be Lindelöf, since the uncountable open cover $\mathcal{C} = \{C\} \cup \{C_a \mid a \in \mathbb{R}\}$ of S^2, where $C = \{(x, y) \in \mathbb{R}^2 \mid x + y < 0\}$ and $C_a = \{(x, y) \in \mathbb{R}^2 \mid a \leq x \text{ and } -a \leq y\}$, contains no proper subcover of S^2. (Draw a picture to see this.) Thus, by (1), S is not Lindelöf. To show that (2) holds it suffices to demonstrate that under **CC**($\mathbb{R}$) S is Lindelöf. For this purpose we will show first that:
(A) $|\sigma| \leq |\mathbb{R}|$
Since $(\mathbb{R}, \tau)$ is second countable it follows immediately that
(B) $|\tau| \leq 2^{\aleph_0}$.
Moreover the set $\mathcal{D}$ of all at most countable subsets of $\mathbb{R}$ satisfies
(C) $|\mathcal{D}| \leq |\mathbb{R}^{\mathbb{N}}| = |\mathbb{R}|^{\mathbb{N}} = (2^{\aleph_0})^{\aleph_0} = 2^{\aleph_0^2} = 2^{\aleph_0}$.
Every $A \in \sigma$ can be decomposed in the form
$A = \text{int}_\tau A \cup (A \setminus \text{int}_\tau A)$, where $A \setminus \text{int}_\tau A$ is easily seen to be at most countable.
Thus (B) and (C) imply:

$$|\sigma| \leq |\tau| \cdot |\mathcal{D}| \leq 2^{\aleph_0} \cdot 2^{\aleph_0} = 2^{\aleph_0 + \aleph_0} = 2^{\aleph_0} = |\mathbb{R}|.$$

Hence (A) holds. This fact implies, via **CC**($\mathbb{R}$), that:
CC(σ) : $\prod_{n \in \mathbb{N}} \mathfrak{U}_n \neq \emptyset$ for every sequence of non–empty subsets $\mathfrak{U}_n$ of σ.

Finally let us consider an open cover $\mathfrak{U}$ of S. Then $\mathfrak{W} = \{\text{int}_\tau U \mid U \in \mathfrak{U}\}$ is an open cover of the open subspace $X = \bigcup \mathfrak{W}$ of $(\mathbb{R}, \tau)$. Since X is second countable, Theorem 4.54 implies that there exists a subset $\{V_n \mid n \in \mathbb{N}\}$

of $\mathfrak{W}$ that covers X. By **CC**(σ) there exists a sequence $(U_n)_{n\in\mathbb{N}}$ in $\mathfrak{U}$ with $V_n = \text{int}_\tau U_n$ for each $n \in \mathbb{N}$. Again it is easily seen that $\mathbb{R} \setminus X$ is at most countable. Thus, by **CC**(σ), there exists a subset $\{W_n \mid n \in \mathbb{N}\}$ of $\mathfrak{U}$ that covers $\mathbb{R} \setminus X$. Consequently $\{U_n \mid n \in \mathbb{N}\} \cup \{W_n \mid n \in \mathbb{N}\}$ is an at most countable subcover of $\mathfrak{U}$. Thus S is a Lindelöf space.

(2) $\Rightarrow$ **(1)** Immediate from Theorem 7.2 and the fact that finite products of compact spaces are compact (see Exercises to Section 4.8, E 1).

Theorem 7.7. *Equivalent are:*

1. *Every Lindelöf T_2-space is paracompact.*
2. *Every Lindelöf T_2-space is normal.*
3. *Every Lindelöf T_2-space is regular.*
4. **CC**($\mathbb{R}$) *fails.*

Proof. **(1)** $\Rightarrow$ **(2)** $\Rightarrow$ **(3)** Immediate.

(3) $\Rightarrow$ **(4)** Enrich the canonical topology τ of the reals by adding the set $B = \mathbb{R} \setminus \{\frac{1}{n} \mid n \in \mathbb{N}^+\}$ as an open set, i.e., by considering the topology σ generated by $\tau \cup \{B\}$. Then the space $(\mathbb{R}, \sigma)$ is a non–regular T_2–space. Thus, by (3), $(\mathbb{R}, \sigma)$ fails to be Lindelöf. Since $(\mathbb{R}, \sigma)$ is second countable, Theorem 4.54, implies that **CC**($\mathbb{R}$) fails.

(4) $\Rightarrow$ **(1)** By Theorem 7.2, (4) implies that Lindelöf T_2–spaces are compact, thus paracompact.

Theorem 7.8. [13] *Equivalent are:*

1. *Every totally disconnected Lindelöf T_2–space is zerodimensional.*
2. **CC**($\mathbb{R}$) *fails.*

Proof. **(1)** $\Rightarrow$ **(2)** Consider the topological space (X, τ), whose underlying set is defined by

$$X = \{(x_n) \in \mathbb{Q}^{\mathbb{N}} \mid \sum_{n\in\mathbb{N}} x_n^2 < \infty\}$$

and whose topology τ is induced by the metric d, defined by

$$d\big((x_n), (y_n)\big) = \sqrt{\sum_{n\in\mathbb{N}} (x_n - y_n)^2}.$$

Then (X, τ) is easily seen to be a totally disconnected, second countable T_2–space. However, (X, τ) fails to be zerodimensional. To see this, consider the point $x = (0, 0, 0, \ldots)$ and its neighborhood

$$U = \{y \in X \mid d(x, y) < 1\}.$$

We will show that there is no clopen neighborhood V of x with $V \subseteq U$. Let V be a neighborhood of x with $V \subseteq U$. Via recursion we will construct an

[13] [Erd40], [Her2002].

element $x = (x_n)$ of X with the following property:

$$\text{(P)} \quad \forall_{n \in \mathbb{N}} \; y_n = (x_0, x_1, \ldots, x_n, 0, 0, \ldots) \in V \text{ and } \mathrm{dist}(y_n, X \setminus V) < \tfrac{1}{2^n}$$

as follows:

1. $x_0 = 0$
2. Let $(x_0, \ldots, x_n)$ be defined such that the corresponding point $y_n = (x_0, \ldots, x_n, 0, 0, \ldots)$ satisfies condition P. Define x_{n+1} to be the smallest element of the set, consisting of all fractions $\frac{m}{2^{n+1}}$ such that:
   a) $m \in \{0, 1, \ldots, 2^{n+1}\}$.
   b) $(x_0, \ldots, x_n, \frac{m}{2^{n+1}}, 0, 0, \ldots) \in V$.
   c) $(x_0, \ldots, x_n, \frac{m+1}{2^{n+1}}, 0, 0, \ldots) \notin V$.

Then x is an element of X that belongs to the closure of V and to the closure of $(X \setminus V)$. Thus V is not clopen.

By (1), the above implies that (X, τ) is not Lindelöf. Thus, by Theorem 4.54, **CC**($\mathbb{R}$) fails.

(2) $\Rightarrow$ **(1)** If **CC**($\mathbb{R}$) fails then, by Theorem 7.2, every totally disconnected Lindelöf T_2–space is compact and thus zerodimensional.

Exercises to Section 7.1:

E 1. [14] Show the equivalence of the following statements:
(1) All *σ–compact* spaces, (i.e., spaces that are countable unions of compact spaces) are Lindelöf.
(2) **CC**.

E 2. [15] Let $\mathbb{N}_l$ be the space of natural numbers with the lower topology (i.e., $A \subseteq \mathbb{N}$ is open in $\mathbb{N}_l$ iff $m \le n \in A$ implies $m \in A$). Show that
(1) $\mathbb{N}_l$ is a Lindelöf T_0–space.
(2) $\mathbb{N}_l^{\mathbb{R}}$ fails to be Lindelöf.

E 3. [16] Show the equivalence of the following conditions:
(1) Finite sums of Lindelöf T_1–spaces are Lindelöf.
(2) Products of Lindelöf T_1–spaces with compact T_1–spaces are Lindelöf.
(3) **CC**($\mathbb{R}$) implies **CC**.

E 4. [17] Show that every unbounded Lindelöf subspace of $\mathbb{R}$ contains an unbounded sequence. (Contrast this with Theorem 3.8.)

---

[14] [Bru82]
[15] [Boer2002]
[16] [Her2002]
[17] [Gut2003]

7.2 Measurability (The Axiom of Determinateness)

Why were set theorists so drawn to study this axiom [of determinateness], *drawn, in fact, to the point where it became the key area of research for all but a few of the best in the field?*

E.M. Kleinberg[18]

If a model of **ZF** *satisfies* **AD**, *then this model is closer to physical reality than any model of* **ZFC**. *For example, the Banach–Tarski paradoxical decomposition of a ball is impossible.*

V.W. Marek and J. Mycielski[19]

Among all alternatives to the axiom of choice **AC** *the axiom of determinateness* **AD** *undoubtedly is the most interesting.*

U. Felgner and K. Schulz[20]

As we have seen in previous sections the Axiom of Choice is not only responsible for some beautiful theorems but also for the creation of some unwelcome monsters, like non–measurable sets of reals. *"It was felt that these choice–generated oddities should not appear among the simpler sets, that they should probably not be definable at all."*[21]

In 1962 Mycielski and Steinhaus[22] introduced a new axiom, the **Axiom of Determinateness**, **AD**, which stipulated that certain infinite, deterministic 2-person games with complete information (cf. Section 6.1) are determinate, i.e., that one of the players has a winning strategy. The authors did not claim this new axiom to be *intuitively true*, but stated that the purpose of their paper is *"only to propose another theory which seems very interesting although its consistency is problematic."*[23] The consistency problem (i.e., the question whether **ZF** + **AD** is consistent, provided **ZF** is) is still unsettled[24]. But set theorists are fairly convinced that **AD** is relatively consistent. In fact they have shown that **ZF** + **AD** is consistent iff **ZFC** and the assumption that infinitely many Woodin cardinals exist, is consistent; and the existence of such large cardinals is one of set theorists' pet beliefs. Though **AD** is incompatible

---

18 [Kle77]
19 [MaMy2001]
20 [FeSch84]
21 [Mad88a]
22 [MySt62]
23 [MySt62]
24 However consistency of **ZF** + **AD** is known to imply consistency of **ZF** + **AD** + **DC** as well as of **ZF** + **AD**+ not **CC**. See [Kec84] and Model A1.

with **AC**, as Theorem 6.5 shows, its consequences are amazing. There are highly desirable results, e.g., the theorem that all sets of real numbers are Lebesgue–measurable. Moreover, there are deep and surprising connections between determinateness of games and the theory of large cardinals. It is the *"richness and internal harmony of these consequences"*[25] that caused the axiom of determinateness *"to have an extraordinary impact on modern set theory"*[26] and has led to a *"very rich and intriguing theory."*[27] Naturally, this theory cannot be presented here. We restrict our attention to outlining the basics and stating some of the consequences without proofs.

Definition 7.9. *Let X be a non–empty set and A a subset of $X^{\mathbb{N}}$. The* game *$G(X, A)$ is played as follows:*

Two players choose alternately consecutive elements $x_0, x_1, x_2, \ldots$ in X, such that each player knows, (besides X and A), whenever it is his term, the tuple of previously choosen elements. The first player (i.e., the one choosing $x_0, x_2, x_4, \ldots$) wins if the resulting sequence (x_n) belongs to A. Otherwise the second player (i.e., the one choosing $x_1, x_3, x_5, \ldots$) wins.

The game $G(X, A)$ is called determinate *provided that one of the players has a winning strategy (see Definition 6.1).*

Recall that $2 = \{0, 1\}$.

Definition 7.10. *Let A be a subset of the unit interval $[0, 1]$. The* game *$G(A)$ is played the same way as the game $G(2, A)$. However, the first player wins if $\sum_{n\in\mathbb{N}} \frac{x_n}{2^{n+1}} \in A$. Otherwise the second player wins.*

The game $G(A)$ is determinate *provided that one of the players has a winning strategy.*

Proposition 7.11. [28] *Equivalent are:*

1. *For each subset A of $\mathbb{N}^{\mathbb{N}}$, the game $G(\mathbb{N}, A)$ is determinate.*
2. *For each subset A of $2^{\mathbb{N}}$, the game $G(2, A)$ is determinate.*
3. *For each subset A of $[0, 1]$, the game $G(A)$ is determinate.*

Proof. **(1)** $\Rightarrow$ **(2)** Let A be a subset of $2^{\mathbb{N}}$. Define $B = \{(x_n) \in \mathbb{N}^{\mathbb{N}} \mid (x_n) \in A \text{ or } \exists n\ x_{2n+1} \notin 2\}$. Then — for either player — a winning strategy for the game $G(\mathbb{N}, B)$ provides easily a wining strategy for $G(2, A)$.

(2) $\Rightarrow$ **(3)** Let A be a subset of $[0, 1]$. Consider the map $f\colon 2^{\mathbb{N}} \to [0, 1]$, defined by $f(x_n) = \sum_{n\in\mathbb{N}} \frac{x_n}{2^{n+1}}$. Then — for either player — a winning strategy for the game $G(2, f^{-1}[A])$ is a winning strategy for $G(A)$.

[25] Cited from [Mad88a].
[26] [Kle77]
[27] [Jen98]
[28] [Myc64]

(3) $\Rightarrow$ (2) Since the map f, defined above, is not injective, we need to apply a small trick. Let A be a subset of $2^{\mathbb{N}}$. Consider the set $M = \{(x_n) \in 2^{\mathbb{N}} \mid \forall_n\, x_{6n} = 0 \text{ and } x_{6n+3} = 1\}$. Then the map $g\colon M \to 2^{\mathbb{N}}$, defined by

$$g((x_n)) = (y_n) \text{ with } \left\{\begin{array}{ll} y_{2n} & = x_{3n+1} \\ y_{2n+1} & = x_{3n+2} \end{array}\right\},$$

is a bijection. Consider the set

$$B = g^{-1}[A] \cup \{(x_n) \in 2^{\mathbb{N}} \mid (x_n) \text{ is not finally constant and } \exists_m\, \forall_k \le m\; x_{6k} = x_{6m+3} = 0\}.$$

As can be seen easily, determinateness of the game $G(2, B)$ implies determinateness of the original game $G(2, A)$. So it suffices to verify the former. This follows immediately from the fact that B is saturated with respect to the map $f\colon 2^{\mathbb{N}} \to [0,1]$ defined above (i.e., from the equation $f^{-1}[f[B]] = B$) and the fact that, by (2), the game $G(f(B))$ is determinate.

(2) $\Rightarrow$ (1) Let A be a subset of $\mathbb{N}^{\mathbb{N}}$. Consider the following partition $\{X, Y, Z\}$ of $\mathbf{2}^{\mathbb{N}}$:

$$X = \{(x_n) \in \mathbf{2}^{\mathbb{N}} \mid \{n \in \mathbb{N} \mid x_{2n} = 0\} \text{ is finite}\},$$
$$Y = \{(x_n) \in (\mathbf{2}^{\mathbb{N}} \setminus X) \mid \{n \in \mathbb{N} \mid x_{2n+1} = 0\} \text{ is finite}\},$$
$$Z = \mathbf{2}^{\mathbb{N}} \setminus (X \cup Y).$$

Consider further the bijection $g\colon Z \to \mathbb{N}^{\mathbb{N}}$, defined as follows:

$y_0 = \min\{k \in \mathbb{N} \mid x_{2k} = 0\}$, i.e., y_0 is the number of consecutive choices of 1's by the first player at the beginning of the game.

$y_1 = \min\{k \in \mathbb{N} \mid x_{2y_0+2k+1} = 0\}$, i.e., y_1 is the number of consecutive choices of 1's by the second player after the first choice of 0 by the first player.

$y_2 = \min\{k \in \mathbb{N} \mid x_{2y_0+2y_1+2(k+1)} = 0\}$,

etc.

Define $B = (g^{-1}[A] \cup Y) \setminus X$. Then it is seen easily that — for either player — a winning strategy for the game $G(2, B)$ provides a winning strategy for the game $G(\mathbb{N}, A)$.

Definition 7.12. AD, *the* Axiom of Determinateness, *states that the equivalent conditions of the preceding proposition are satisfied.*

An impressive consequence of the Axiom of Determinateness is the following result that we present without proof:

Theorem 7.13. [29] *Under* **AD** *every subset of* $\mathbb{R}$ *is Lebesgue–measurable.*

The above results indicates that Lebesgue measure for subsets of $\mathbb{R}$ is better behaved under **AD** than under **AC**. However, it is known that in **ZF** Lebesgue measure may fail to be σ–additive. So immediately the question pops up whether by gaining something (the measurability of all subsets of $\mathbb{R}$) one looses something else (the σ–additivity of Lebesgue measure). Fortunately this is not so:

---

[29] [MySw64]. For improved proofs see, e.g., [Jec73], [Kle77] or [Mar2003].

Proposition 7.14. [30] *Under* **AD** *Lebesgue measure is* σ*–additive.*

Proof. The familiar proof of σ–additivity of the Lebesgue measure uses, in several places, the following weak choice principle, where τ denotes the topology of $\mathbb{R}^n$:

CC(τ): For each sequence (X_n) of non–empty subsets X_n of τ, the product $\prod_{n\in\mathbb{N}} X_n$ is not empty.

The fact that τ has a countable base immediately implies $|\tau| = 2^{\aleph_0} = |\mathbb{R}|$. Thus **CC**($\tau$) is equivalent to **CC**($\mathbb{R}$). Thus **CC**(τ), and hence the σ–additivity of Lebesgue measure follows from **AD** via the next proposition.

Proposition 7.15. [31] *Under* **AD**

1. **CC**($\mathbb{R}$) *holds,*
2. **AC**($\mathbb{R}$) *fails.*

Proof. (**1**) Consider $M = \{2n+1 \mid n \in \mathbb{N}\}$. Then $|\mathbb{N}^M| = 2^{\aleph_0} = |\mathbb{R}|$ implies that it suffices to show that $\prod_{n\in\mathbb{N}} A_n \neq \emptyset$ for each sequence (A_n) of non–empty subsets A_n of $\mathbb{N}^M$. Consider the game $G(\mathbb{N}, A)$, where $A = \{(x_n) \in \mathbb{N}^\mathbb{N} \mid \forall n \in \mathbb{N}\ (x_{2n+1}) \notin A_{x_0}\}$. Since the sets A_n are non–empty the first player has no winning strategy. Thus, by **AD**, the second player has a winning strategy. This provides him with a function that associates with each strategy σ_m of the first player, of the form play "$x_0 = m$ and $x_{2n} = 0$ for $n \geq 1$", a sequence $s_m = (x_{2n+1})$ in A_m. Thus $(s_m) \in \prod_{m\in\mathbb{N}} A_m$.

(**2**) In Theorem 6.5 it has been shown that **AC** implies the existence of a subset A of $\mathbf{2}^\mathbb{N}$ such that the game $G(2, A)$ is not determinate. A straightforward analysis of the proof reveals that only **AC**($\mathbb{R}$) is needed. Thus **AC**($\mathbb{R}$) implies the failure of **AD**. By contraposition **AD** implies the failure of **AC**($\mathbb{R}$). An alternative proof can be obtained by means of theorem 7.13 via the construction 5.7 of non–measurable Vitali Monster.

Proposition 7.16. [32] *Under* **AD** *there is no free ultrafilter on* $\mathbb{N}$.

Proof. This follows immediately from Theorem 7.13 via the construction 5.9 of non–measurable Sierpiński Monsters by means of **WUT**($\mathbb{N}$).

---

[30] [Myc64]
[31] [Myc64]
[32] [Myc64]. Note, however, that there are even σ–complete free ultrafilters on $\aleph_1$. [Mig81], [Ver94].

Corollary 7.17. *Under* **AD** *the discrete space of natural numbers is ultrafilter–compact and Tychonoff–compact.*

Proof. Immediate from Proposition 7.16 and Theorem 3.32(2).

Proposition 7.18. *Under* **AD** *all solutions of the Cauchy–equation are continuous.*

Proof. Immediate from Theorems 7.13 and 5.5 in view of Exercises to Section 5.1, E 7.

Corollary 7.19. [33] *Under* **AD** *the additive group* $\mathbb{R}$ *has no non–trivial direct summand.*

Proof. Assume that $\mathbb{R}$ can be expressed as a direct sum $A \oplus B$ of two non–zero subgroups. Then the map $f\colon \mathbb{R} \to \mathbb{R}$, defined by $f(a+b) = a$ for $a \in A$ and $b = B$, is a non–continuous solution of the equation $f(x+y) = f(x) + f(y)$, contradicting Proposition 7.18.

Corollary 7.20. [34] *Under* **AD**, $\mathbb{R}$ *considered as vector space over the field* $\mathbb{Q}$ *has no basis.*

Proof. If B would be a basis of the vector space $\mathbb{R}$ over $\mathbb{Q}$, then for any $b \in B$ the set $\mathbb{Q} \cdot b = \{r \cdot b \mid r \in \mathbb{Q}\}$ would be a non–trivial direct summand of the additive group of $\mathbb{R}$, contradicting Corollary 7.19.

Theorem 7.21. [35] *Consider* $K = X/\varrho$ *where* $X = \{0,1\}^{\mathbb{N}}$ *and* ϱ *is the equivalence relation on* X*, defined by*

$$(x_n)\varrho(y_n) \text{ iff } \{n \in \mathbb{N} \mid x_n \neq y_n\} \text{ is finite.}$$

Under **AD** *the following hold:*

1. *There exists a family* $(F_i)_{i\in I}$ *of 2–element subsets* F_i *of* K *with* $\prod_{i\in I} F_i = \emptyset$.
2. K *is not linearly orderable.*
3. $|\mathbb{R}| < |K|$ *and* $|K| <^* |\mathbb{R}|$.

Proof. (1) For each $x = (x_n)$ in X define $x^* = (1 - x_n)$. Denote further, for each x in X, its equivalence class with respect to ϱ by $[x]_\varrho$. Then the

---

[33] [FeSch84]

[34] [Sie27], [Sie30], [Sie58, p. 77], [Myc64]. Observe also that under **AD** the factor group $\mathbb{R}/\mathbb{Q}$ of the additive group $\mathbb{R}$ fails to be linearly orderable. See [FeSch84].

[35] [Myc64]

set $\{\{[x]_\varrho, [x^*]_\varrho\} \mid x \in X\}$ of 2–element subsets of K, considered as family indexed by itself, has an empty product, since otherwise (see Exercises to Section 5.1, E 1) there would exist a non–measurable subset of $\mathbb{R}$, contradicting Theorem 7.13.

(**2**) Immediate from (1) and Exercises to Section 1.1, E 2 (8).

(**3**) Observe first that $|K| \neq |\mathbb{R}|$, since $\mathbb{R}$ is linearly orderable but, by (2), K fails to be linearly orderable. Observe next that $|K| \leq^* |\mathbb{R}|$, since there exists a bijection $b\colon \mathbb{R} \to X$, and a surjection $p\colon X \to K$, defined by $p(x) = [x]_\varrho$. Thus it remains to be shown that $|\mathbb{R}| \leq |K|$. Here the crucial observation, due to Sierpiński[36], is the fact that the function $f\colon (0,1) \to \mathcal{P}(\mathbb{N}^2)$, defined by $f(x) = \{(n, \mathrm{int}(n \cdot x)) \mid n \in \mathbb{N}\}$, where

$$\mathrm{int}(a) = \max\{n \in \mathbb{N} \mid n \leq a\},$$

has the following property:

($\bullet$) If $x \neq y$, then $f(x) \cap f(y)$ is finite.

To see this, let x and y be different elements of $(0,1)$, and let $(n, \mathrm{int}(nx)) = (m, \mathrm{int}(my))$ be an element of $f(x) \cap f(y)$. Then $n = m$ and $\mathrm{int}(nx) = \mathrm{int}(ny)$. The latter implies $|nx - ny| < 1$ and thus $n < \frac{1}{|x-y|}$. Thus $f(x) \cap f(y)$ has at most $\mathrm{int}\left(\frac{1}{|x-y|}\right)$ members. Let $g\colon \mathbb{R} \to (0,1)$ and $h\colon \mathbb{N}^2 \to \mathbb{N}$ be arbitrary bijections and let $\chi\colon \mathcal{P}(\mathbb{N}) \to X$ be the map associating with any $A \subseteq \mathbb{N}$ its characteristic function χ_A. Then the map $k = \chi \circ \mathcal{P}(h) \circ f \circ g\colon \mathbb{R} \to X$ has the property ($\bullet$) and thus the map $p \circ k\colon \mathbb{R} \to K$ is injective.

Finally, let us mention without proof the following remarkable results concerning the Continuum Hypothesis.

Theorem 7.22. [37] *Under* **AC** *the following hold:*

1. $2^{\aleph_0}$ *and* $\aleph_1$ *are w.r.t.* $\leq$ *incomparable minimal successors of* $\aleph_0$. *Thus*
   - **CH**, *the Continuum Hypothesis, holds.*
   - **AH**(0), *the Special Aleph–Hypothesis, fails.*
2. $\aleph_0 <^* \aleph_1 <^* 2^{\aleph_0}$.
   Thus, w.r.t. $\leq^*$ *the Continuum Hypothesis fails.*
3. *W.r.t.* $\leq$ *there are at least 3 cardinals between* $2^{\aleph_0}$ *and* $2^{(2^{\aleph_0})}$*:*
   $2^{\aleph_0} < (2^{\aleph_0} + \aleph_1) < 2^{\aleph_1} < (2^{\aleph_1} + |K|) < 2^{(2^{\aleph_0})}$
   (where K *is defined in Theorem 7.21).*

---

36 [Sie58, p. 77]

37 [Myc64]

The following diagram summarizes results from Sections 4.11 to 7.2 concerning the production of mathematical monsters:

Diagram 7.23.

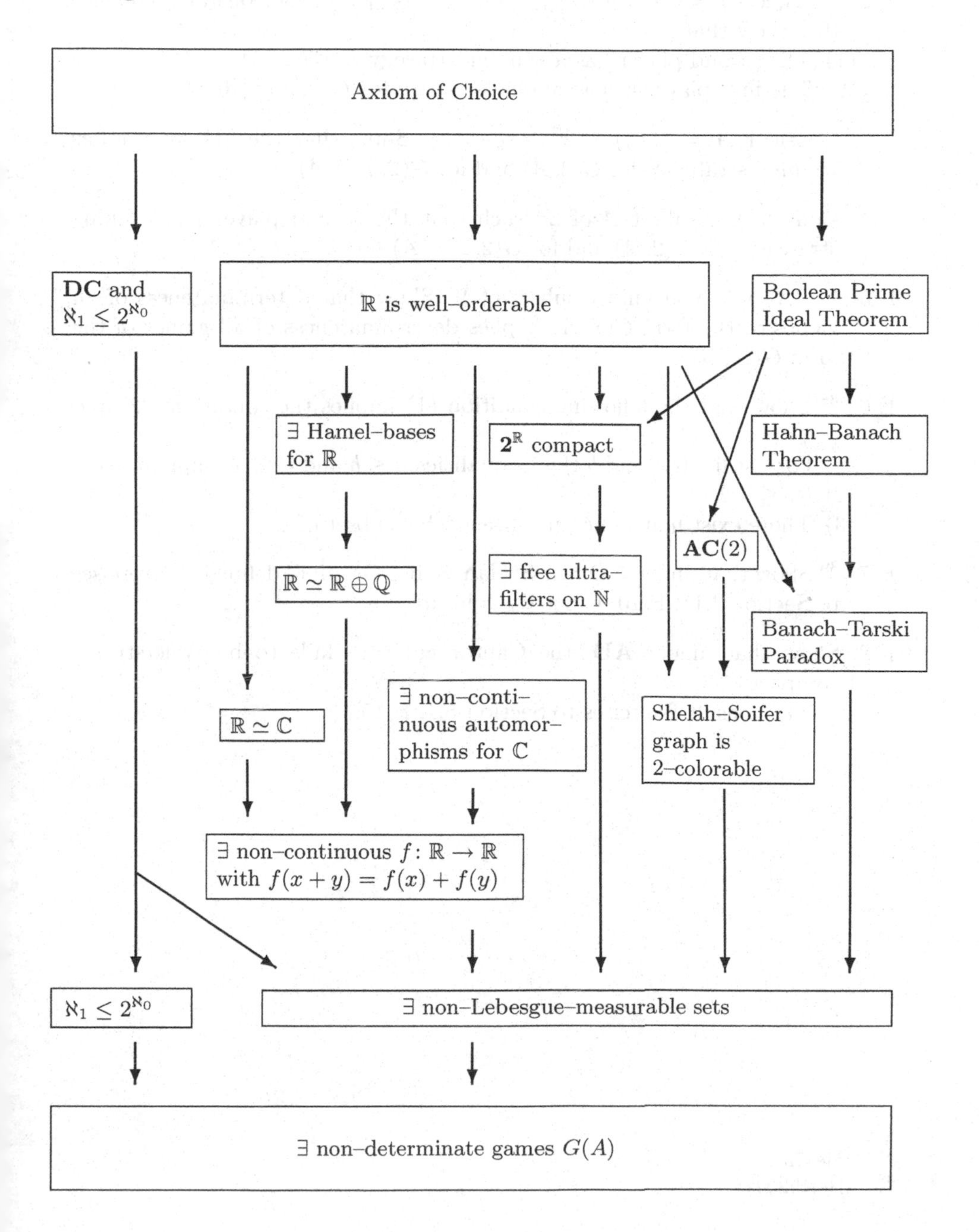

Exercises to Section 7.2:

E 1. Show that, for every countable (resp. cocountable) subset A of $2^{\mathbb{N}}$, the second (resp. first) player has a winning strategy for $G(2, A)$.

E 2. Consider $A = \{(x_n) \in 2^{\mathbb{N}} \mid \forall_n \ x_{2n+1} = x_{2n}\}$ and the constant sequence (0). Show that:
(1) The second player has a winning strategy for $G(2, A)$.
(2) The first player has a winning strategy for $G(2, A \cup \{(0)\})$.

E 3. Consider $A = \{(x_n) \in 2^{\mathbb{N}} \mid x_0 = 0\}$. Show that the first player has winning strategies for $G(2, A)$ and for $G(2, 2^{\mathbb{N}} \setminus A)$.

E 4. Construct a subset A of $2^{\mathbb{N}}$ such that the second player has winning strategies for $G(2, A)$ and for $G(2, 2^{\mathbb{N}} \setminus A)$.

E 5. Let X be a non–empty subset of Y. Show that determinateness of all games of the form $G(Y, A)$ implies determinateness of all games of the form $G(X, B)$.

E 6. [38] Show that the following condition (1) implies the conditions (2) and (3):
(1) For cardinals a and b the inequalities $a \leq b$ and $b \leq^* a$ imply $a = b$.
(2) $\aleph_1 \leq 2^{\aleph_0}$.
(3) There exist non–Lebesgue–measurable subset of $\mathbb{R}$.

E 7. [39] Show that under **AD**, the Shelah–Soifer Graph G defined in Exercises to Section 4.11, E 10, is not $\aleph_0$–colorable.

E 8. Show that, under **AD**, the Cantor cube $\mathbf{2}^{\mathbb{R}}$ fails to be Weierstrass–compact.
[Hint: Consult Exercises to Section 5.1, E 15.]

[38] [Sie47a]
[39] [HeRh2005]

Appendix: Models

In the main text several **ZF**–models and some of their properties have been mentioned. In this Appendix we illustrate properties of some of these models by means of diagrams, where for a model M

- P means that M has property P.
- P means that M fails to satisfy P.
- P means that we do not know whether M satisfies P.

Most of the data presented here are taken from [HoRu98] where these and many other models are described and analyzed most thoroughly and in great detail.

A.1 **AD** and **DC**

Though no models for **AD** have been constructed so far, it is known[40] that if there exists a **ZF**–model that satisfies **AD**, then there also exists a **ZF**–model that satisfies **AD** and **DC**. For properties of any such model see Diagram A.1.

Note in particular that **DC** ensures the validity of

- most results from elementary analysis (Section 4.6),
- σ–additivity of Lebesgue–measure (Exercises to Section 5.1, E 13 resp. Proposition 7.14),
- the Baire Category Theorem for complete metric and for compact Hausdorff spaces (Theorem 4.106).

Moreover, **AD** implies that

- all subsets of $\mathbb{R}$ (resp. of $\mathbb{R}^n$) are Lebesgue–measurable (Theorem 7.13),
- all real solutions of the Cauchy–equation $f(x+y) = f(x) + f(y)$ are continuous (Proposition 7.18),
- no paradoxical decompositions exist (Section 5.2).

On the other hand:

[40] see [Kec84]

- **WUF**($\mathbb{N}$) fails (Proposition 7.16), hence
  - the discrete space $\mathbb{N}$ is ultrafilter–compact and Tychonoff compact (Theorem 3.32),
  - the Baire Category Theorem fails for ultrafilter–compact T_3–spaces (Theorem 4.108).
- **PIT** fails, thus
  - the Tychonoff Theorem fails even for Cantor cubes $\mathbf{2}^I$ (Theorem 4.70); moreover $\mathbf{2}^{\mathbb{R}}$ fails to be Weierstrass–compact (Exercises to Section 7.2, E 8),
  - the Ascoli Theorem fails (Theorem 4.91).
- **AC**(2) fails (Exercises to Section 5.1, E 1), thus
  - for every natural number $n \geq 2$ there exists a graph that fails to be n–colorable, though each of its finite subgraphs is n–colorable (Theorems 4.113 and 4.115),
  - $\mathbb{R}$, considered as a vector space over $\mathbb{Q}$, has no basis (Proof of Disaster 5.2 (2)).
  - The Adjoint Functor Theorem fails (Theorem 4.51).

  Finally, since **AC** fails but **WUF**(?) holds:
- The Tychonoff Theorem fails even for ultrafilter–compact spaces (Theorem 4.80).

A.2 Shelah's Second Model[41]

In this model, **DC** holds and any subset of $\mathbb{R}$ is Lebesgue–measurable. As Diagram A.2 shows this model shares many of the features of A.1–models.

A.3 Howard–Rubin's First Model[42]

In this model, **AC**($\mathbb{R}$), **CC**, and **PIT** hold, but **DC** and **KW** fail. See Diagram A.3. Note in particular that here:

- Most results from elementary analysis are valid (Section 4.6)
- The Tychonoff Theorem holds for Hausdorff spaces (Theorem 4.70), and for countable products (Exercises to Section 4.8, E 4), but not in full generality (Theorem 4.68).
- The Čech–Stone Theorem holds (Theorems 3.22 and 4.8).
- The Ascoli Theorem holds (Theorem 4.91).
- The Baire Category Theorem holds for totally bounded or second countable complete metric spaces (Theorem 4.104) and for countable products of compact metric spaces (Theorem 4.105), but fails for complete metric spaces and for compact Hausdorff spaces (Theorem 4.106).
- The Hahn–Banach–Theorem holds (Diagram 5.25).
- Every open lattice has a maximal filter (Theorem 4.36), but not every closed lattice has a maximal filter (Theorem 4.32).

---

[41] M 38 in [HoRu98]. A similar model has been constructed earlier by Solovay [Sol65]. These constructions assume that in some **ZF**–models, inaccessible cardinals exist.

[42] N 38 in [HoRu98].

- For every n, a graph is n–colorable provided that each of its finite subgraphs is n–colorable (Theorems 4.113, 4.115 and Exercises to Section 4.11, E 4).
- Countable sums of normal spaces are normal (Theorem 4.66); however there exists an orderable space that is a sum of normal spaces but fails to be normal itself (Exercises to Section 4.7, E 5).

A.4 Cohen's First Model[43]

In this model **PIT** holds, but **Fin**($\mathbb{R}$) and thus **CC**($\mathbb{R}$) fail. See Diagram A.4.

Note in particular that here:

- Lindelöf = compact for T_1–spaces (Theorem 7.2).
- Finite products of Lindelöf T_1–spaces are Lindelöf (Theorem 7.6).
- Arbitrary products of Lindelöf T_2–spaces are Lindelöf (Theorem 7.4).
- The Čech–Stone Theorem for Lindelöf spaces holds: Lindelöf T_2–spaces form an epireflective subcategory of the category of T_2–spaces (Remark 7.5).
- Lindelöf T_2–spaces are normal (Theorem 7.7).
- Totally disconnected Lindelöf T_2–spaces are zerodimensional (Theorem 7.8).

Moreover, due to **PIT** alone:

- The Tychonoff Theorem holds for Hausdorff spaces (Theorem 4.70).
- The Čech–Stone Theorem holds (Theorems 3.22 and 4.8).
- The Hahn–Banach theorem holds (Diagram 5.25).
- Every open lattice has a maximal filter (Theorem 4.36).
- For every n a graph is n–colorable provided each of its finite subgraphs is n–colorable (Theorems 4.113, 4.115, and Exercises to Section 4.11, E 4).

However, since **Fin**($\mathbb{R}$) fails:

- Many results in elementary analysis fail (Section 4.6).
- $[0,1]$ fails to be Alexandroff–Urysohn compact (Theorem 3.32).
- There exist infinite subsets of $\mathbb{R}$ without any decreasing or increasing sequences (Disaster 4.25).

A.5 Fraenkel's Second Model[44]

In this model **AC**($\mathbb{R}$) holds, but **AC**(2), **CC**(2), and even **PCC**(2) fail. See Diagram A.5.

Observe in particular that here

- The discrete space $\mathbb{N}$ is Lindelöf, but the sum of $\mathbb{N}$ with a suitable compact T_2–space fails to be Lindelöf (Remark 4.63).
- A countable union of pairwise disjoint 2–element sets can fail to be countable, even D–infinite (Proposition 3.6).
- D–finite unions of D–finite sets can be D–infinite (Disaster 4.3).

[43] M1 in [HoRu98].

[44] N2(2) in [HoRu98].

- Images of D–finite sets can be D–infinite (Disaster 4.3).
- The power set of a D–finite set can be D–infinite (Disaster 4.3).
- Countable products of 3–element spaces can fail to be compact (Theorem 4.77) or Baire (Exercises to Section 4.10, E 10).

A.6 **Pincus–Solovay's First Model**[45]

In this model there are no free ultrafilters (i.e., **WUF**(?) fails), but **DC** holds. See Diagram A.6.

So here:

- The Tychonoff Theorem holds for ultrafilter–compact spaces (Theorem 4.80),
  and for countable products of compact spaces (Proposition 4.72),
  but fails for compact Hausdorff spaces (Theorem 4.70).
- The Ascoli Theorem fails (Theorem 4.91).
- The Baire Category Theorem holds for complete metric and for compact Hausdorff spaces (Theorem 4.106), but fails for ultrafilter–compact T_3–spaces (Theorem 4.108).

A.7 **Fraenkel's First Model**[46]

In this model, there exist amorphous sets. See Exercises to Section 4.1, E 11; Section 4.3, E 4 and Section 4.10, E 11.

A.8 **Feferman–Levy Model**[47]

In this model, $\mathbb{R}$ is the countable union of countable sets.

Consequently here:

- $\mathbb{R}$ is a sequential space (Proposition 4.57).
- $\mathbb{R}$ is neither Fréchet nor Lindelöf (Theorem 4.54).
- Lebesgue measure fails to be σ–additive (Exercises to Section 5.1, E 13).

A.9 **Howard–Rubin's Second Model**[48]

In this model, **AC**($\mathbb{R}$), **DC** and **PIT** hold, but **KW** fails.

---

[45] M27 in [HoRu98].
[46] N1 in [HoRu98].
[47] M9 in [HoRu98].
[48] N40 in [HoRu98].

Diagram A.1: ZF-Models satisfying AD and DC

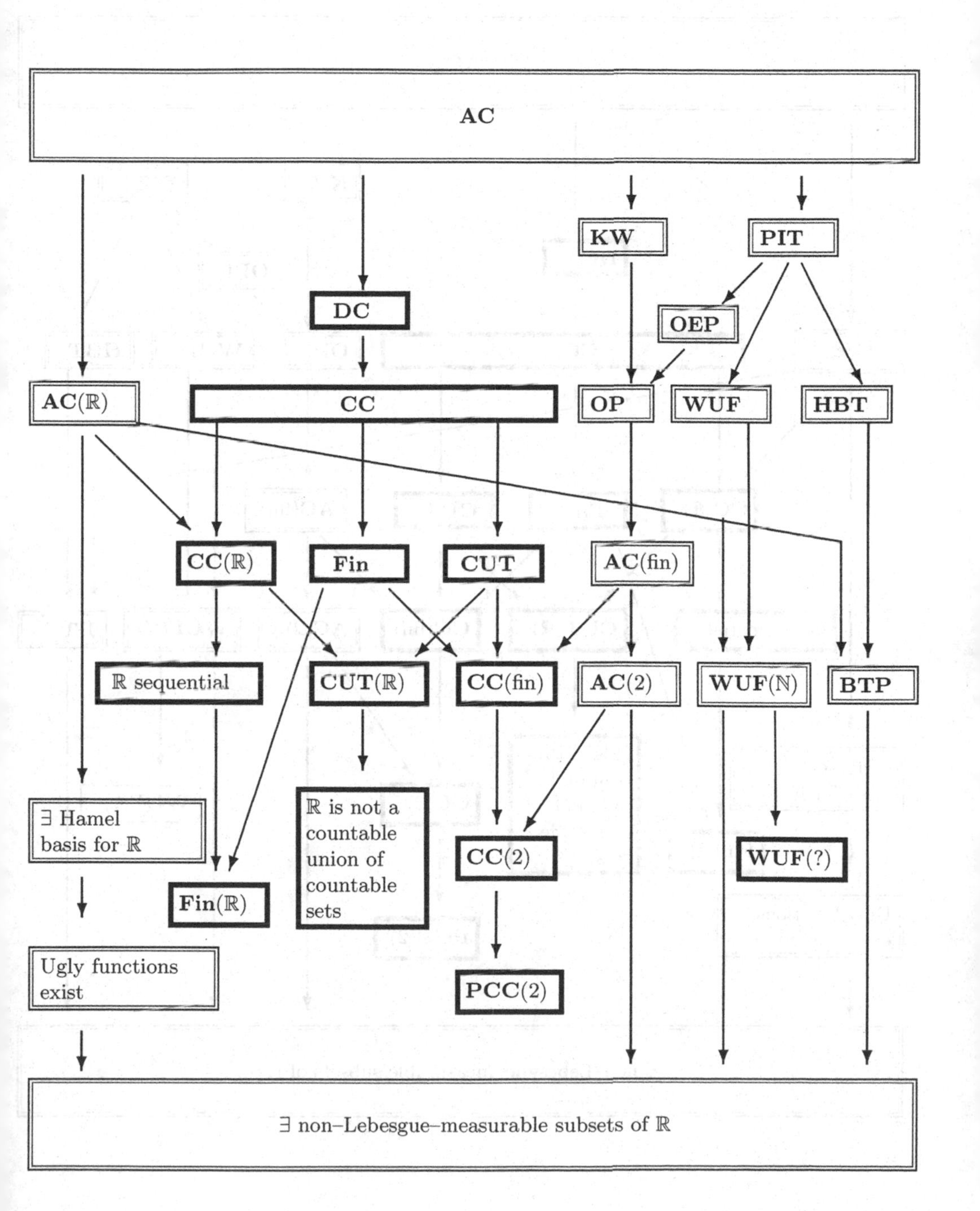

Diagram A.2: Shelah's Second Model

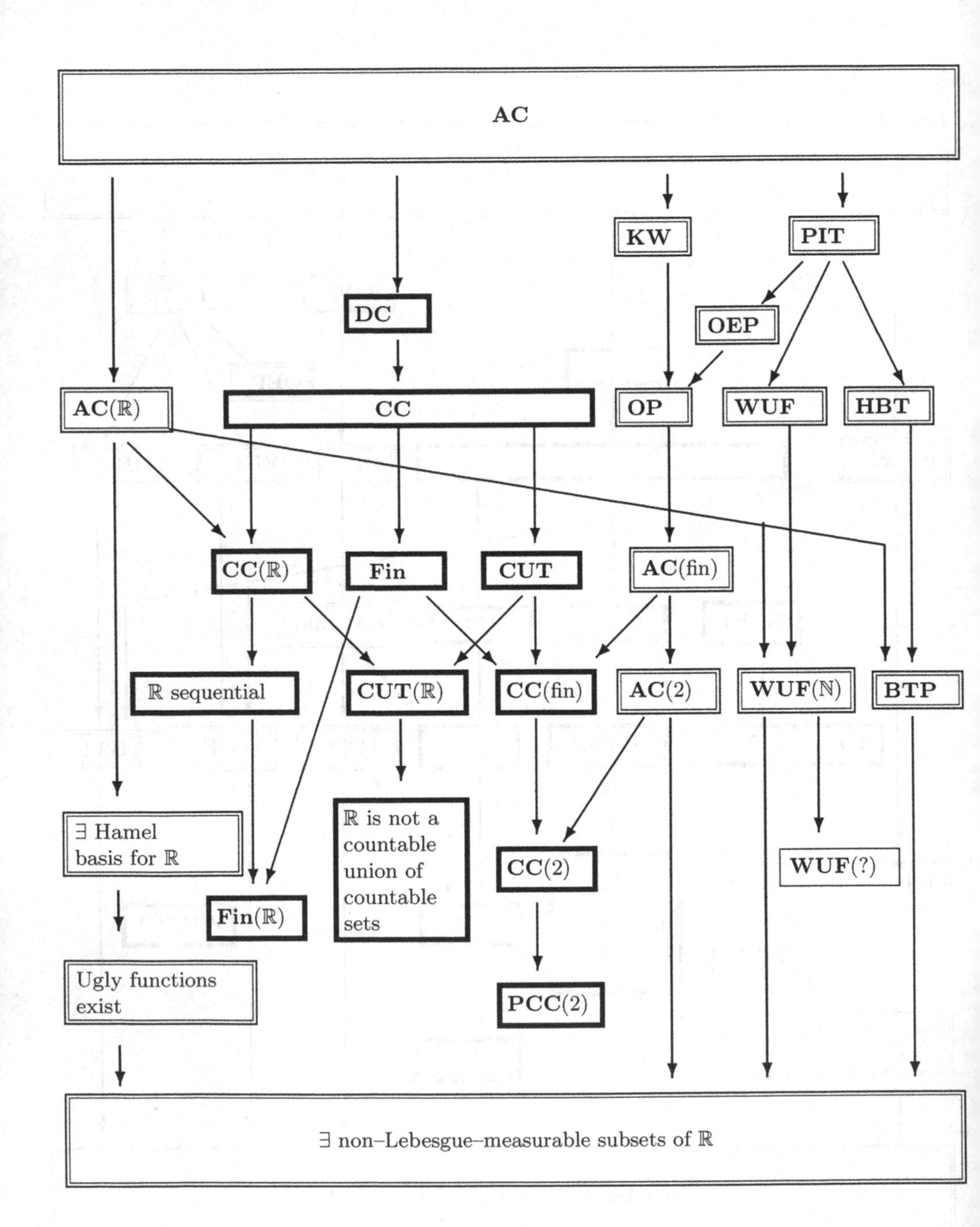

Diagram A.3: Howard–Rubin's First Model

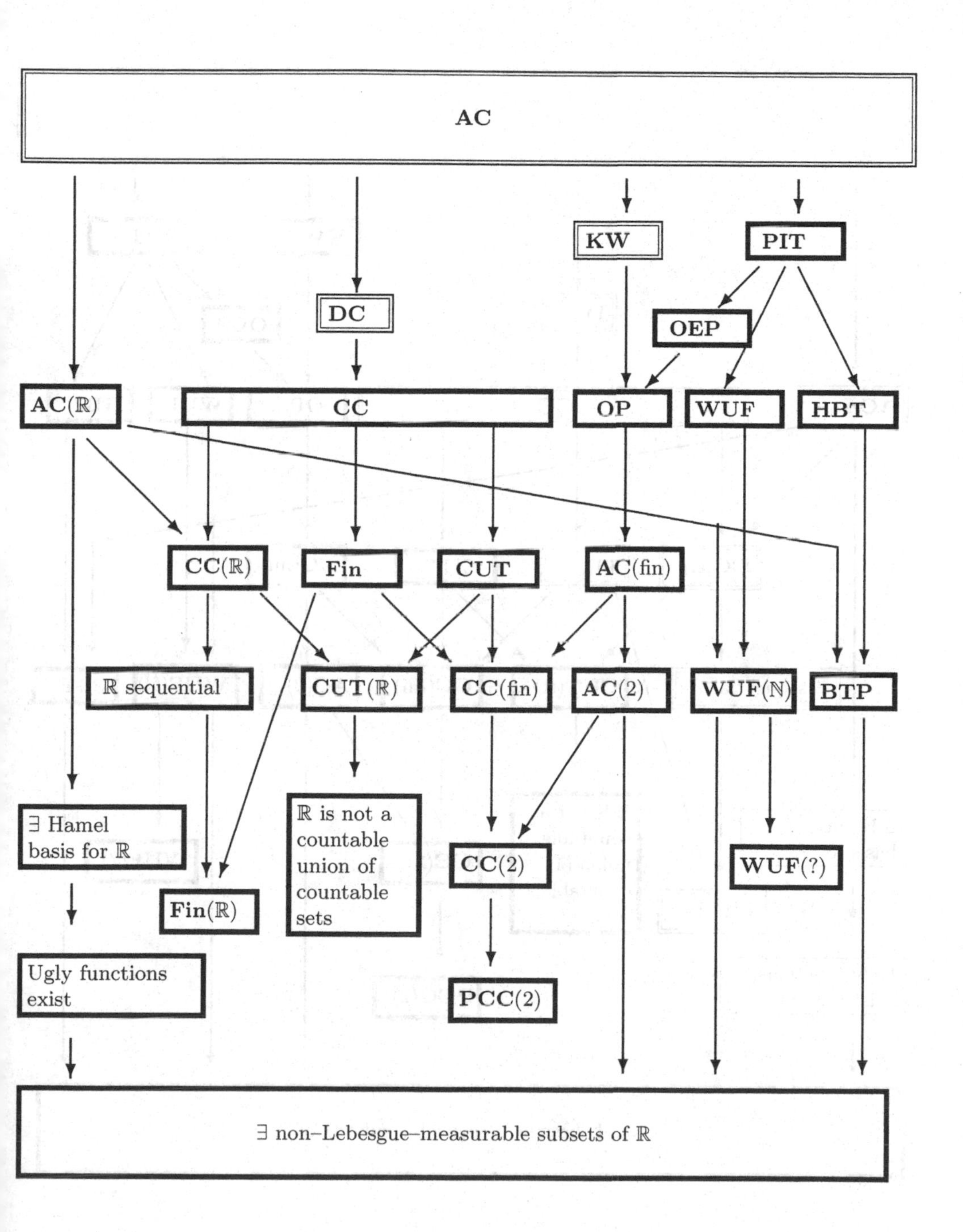

Diagram A.4: Cohen's First Model

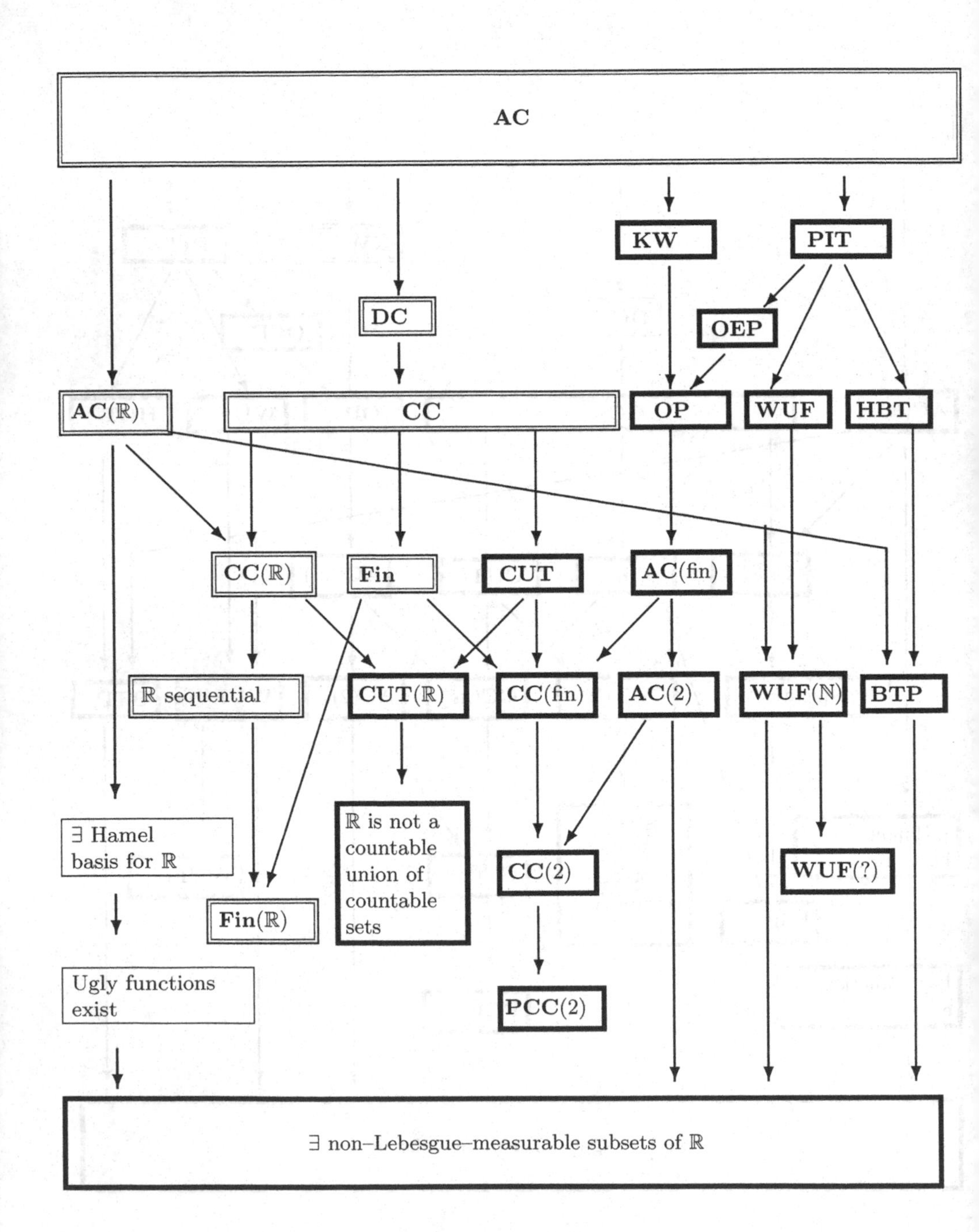

Diagram A.5: Fraenkel's Second Model

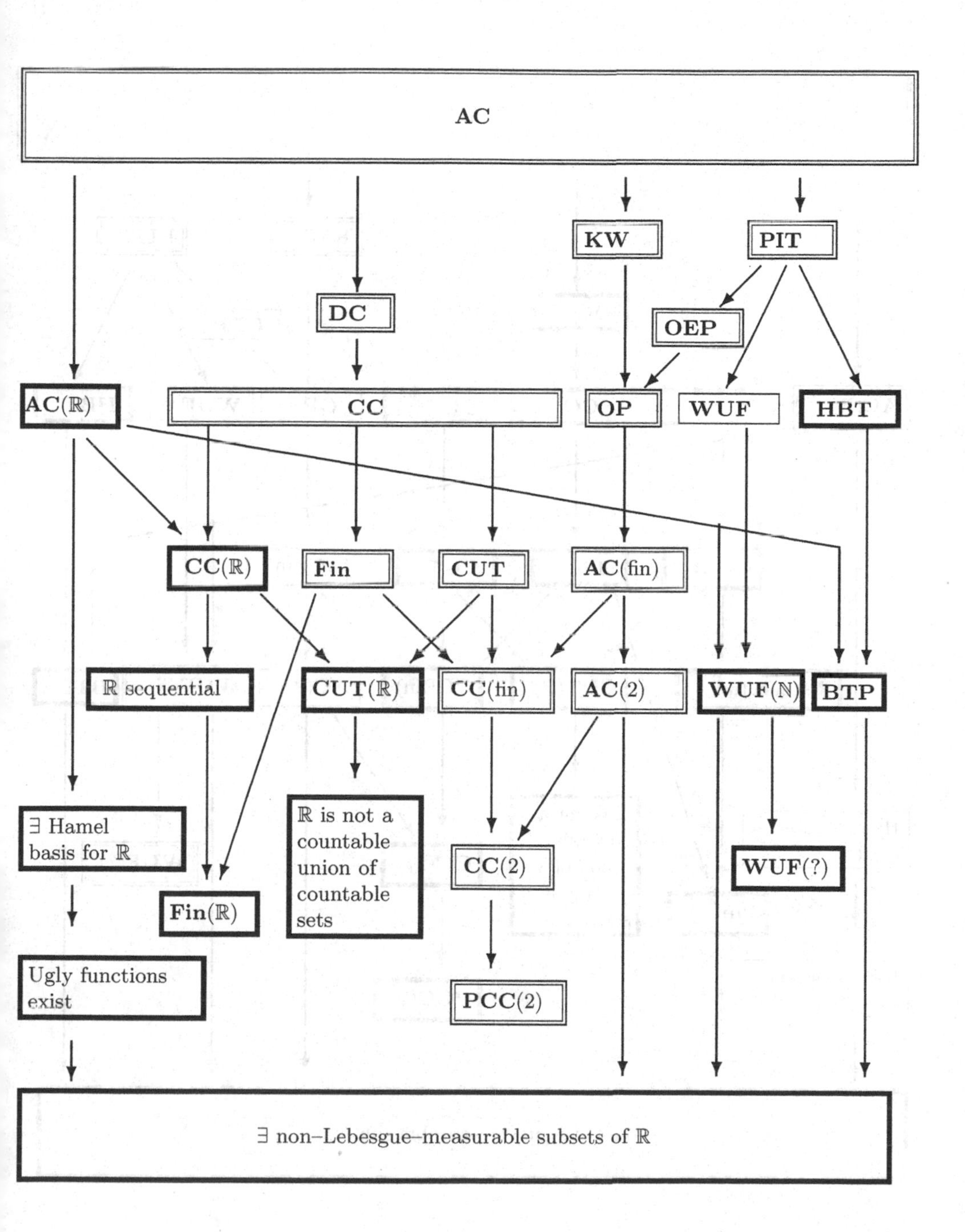

Diagram A.6: Pincus–Solovay's First Model

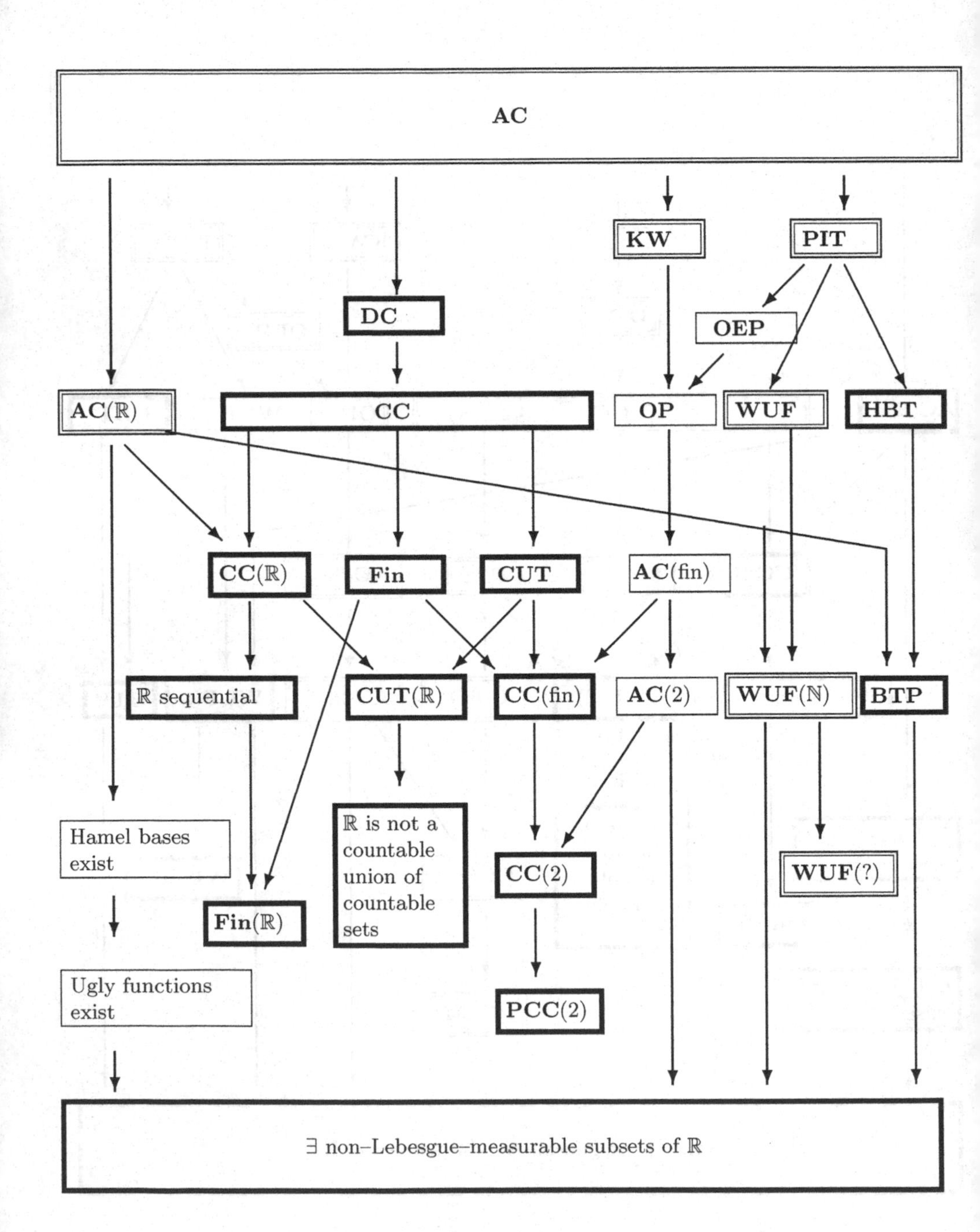

References

[AHS2004] J. Adámek, H. Herrlich and G.E. Strecker. Abstract and Concrete Categories. The Joy of Cats. http://katmat.math.uni-bremen.de/acc/acc.pdf

[AlUr29] P. Alexandroff and P. Urysohn. Mémoire sur les espaces topologiques compacts. *Verh. Nederl. Akad. Wetensch. Afd. Natuurk. Sect. I*, 14:1–96, 1929.

[AnKuNe94] H. Andréka, Á. Kurucz and I. Németi. Connections between axioms of set theory and basic theorems of universal algebra. *J. Symb. Logic*, 59:912–923, 1994.

[AlOr45] A. Alexiewicz and W. Orlicz. Remarque sur l'équation fonctionelle $f(x+y) = f(x) + f(y)$. *Fund. Math.*, 33:314–315, 1945.

[Ash75] L.J. Ash. A consequence of the axiom of choice. *J. Austral. Math. Soc. Ser. A* 19:306–308, 1975.

[Bana20] S. Banach. Sur l'équation fonctionelle $f(x+y) = f(x) + f(y)$. *Fund. Math.*, 1:123, 1920.

[Bana23] S. Banach. Sur le problème de la mesure. *Fund. Math.*, 4:7–33, 1923.

[BaTa24] S. Banach and A. Tarski. Sur la décomposition des ensembles de points in parties respectivement congruents. *Fund. Math.*, 6:244–277, 1924.

[Ban61] B. Banaschewski. On some theorems equivalent with the axiom of choice. *Z. Math. Logik Grundl. Math.*, 7:279–282, 1961.

[Ban79] B. Banaschewski. Compactification and the axiom of choice. Unpublished manuscript, 1979.

[Ban92] B. Banaschewski. Algebraic closure without choice. *Z. Math. Logik Grundl. Math.*, 38:383–385, 1992.

[Ban93] B. Banaschewski. Supercompactness, products and the axiom of choice. *Kyungpook Math. J.*, 33:111–114, 1993.

[Ban94] B. Banaschewski. A new proof that "Krull implies Zorn". *Math. Logic Quart.*, 40:478–480, 1994.

[Ban98] B. Banaschewski. Choice principles and compactness conditions. *Math. Logic Quart.*, 44:427–430, 1998.

[BaBr86] B. Banaschewski and G.C.L. Brümmer. Thoughts on the Cantor–Bernstein Theorem. *Quaest. Math.*, 9:1–27, 1986.

[BaMo90] B. Banaschewski and G.H. Moore. The dual Cantor–Bernstein Theorem and the Partition Principle. *Notre Dame J. Formal Logic*, 31:375–381, 1990.

[BeFr58] P. Bernays and A.A. Fraenkel. Axiomatic Set Theory. North Holland, 1958.

[BeHe98] H.L. Bentley and H. Herrlich. Countable choice and pseudometric spaces. *Topology Appl.*, 85:153–164, 1998.

[BeHe99] H.L. Bentley and H. Herrlich. Compactness and rings of continuous functions – without the axiom of choice. *Proc. Symp. Cat. Topol.* Cape Town 1994, Univ. Cape Town 1999. (Eds. B. Banaschewski, C.R.A. Gilmour, and H. Herrlich), 47–54.

[Ber08] F. Bernstein. Zur Theorie der trigonometrischen Reihe. *Königl. Sächs. Ges. Wis., Leipzig, Math–Phys. Sitzungsbericht*, 1908.

[BeDeSch99] R. Berr, F. Delon and J. Schmid. Ordered fields and the ultrafilter theorem. *Fund. Math.*, 159:231–241, 1999.

[Bla77] C.E. Blair. The Baire category theorem implies the principle of dependent choices. *Bull. Acad. Polon. Sci.*, 25:933–934, 1977.

[Blass77] A. Blass. A model without ultrafilters. *Bull. l'Acad. Polon. Sci. Ser. Sci. Math. Astr. Phys.*, 25:329–331, 1977.

[Blass79] A. Blass. Injectivity, projectivity, and the axiom of choice. *Transactions Amer. Math. Soc.*, 255:31–59, 1979.

[Blass84] A. Blass. Existence of bases implies the axiom of choice. *Contemporary Mathematics*, 31:31–33, 1984.

[Blei64] M.N. Bleicher. Theorems on vector spaces and the axiom of choice. *Fund. Math.*, 54:95–107, 1964.

[Boer2002] R. Börger. On the powers of a Lindelöf space. *Seminarberichte Fb. Math. FernUniv. Hagen* 73:1–2, 2002.

[Boh49] N. Bohr. Discussions with Albert Einstein on epistemological problems in atomic physics. In: Albert Einstein, Philosopher–Scientist. Library of living philosophers. Evanston, Illinois 1949.

[Bor14] E. Borel. *Leçons sur la théorie des fonctions.* Paris, 1914.

[BrCe75] A.M. Bruckner and J. Ceder. On improving Lebesgue measure. *Nordisk Matem. Tidskrift*, 23:59–68, 1975.

[BrEr51] N.G. de Bruijn and P. Erdös. A colour problem for infinite graphs and a problem in the theory of relations. *Indag. Math.*, 13:371–373, 1951.

[Bru82] N. Brunner. σ–kompakte Räume. *manuscr. math.*, 38:375–379, 1982.

[Bru82a] N. Brunner. Lindelöf Räume und Auswahlaxiom. *Anz. Österr. Akad. Wiss., Math.-Naturwiss. Kl.*, 119:161–165, 1982.

[Bru82b] N. Brunner. Dedekind–Endlichkeit und Wohlordenbarkeit. *Monatshefte Math.*, 94:9–31, 1982.

[Bru83] N. Brunner. Kategoriesätze und Multiples Auswahlaxiom. *Z. Math. Logik Grundlagen Math.*, 29:435–443, 1983.

[Bru83a] N. Brunner. The axiom of choice in topology. *Notre Dame J. Formal Logic*, 24:305–317, 1983.

[Bru83b] N. Brunner. Folgenkompaktheit und Auswahlaxiom. *Archiv Math. 3, Ser. Fac. Sci. Nat. UJEP Brunensis*, 19:143–144, 1983.

[Bru84] N. Brunner. Amorphe Potenzen kompakter Räume. *Archiv Math. Logik Grundlagenforschung*, 24:119–135, 1984.

[Bru86] N. Brunner. Ultraproducts and the axiom of choice. *Archiv Math. Brno*, 22:175–180, 1986.

[Bru88] N. Brunner. Mathematische Intuition, Kontinuumshypothese und Auswahlaxiom. *Jahrbuch Kurt Gödel–Ges.*, 96–101, 1988.

[Bru2001] N. Brunner. Maximal ideals and the axiom of choice. *Unsolved problems in mathematics for the 21st century.* Amsterdam 105:183–192, 2001.

[BrHo92] N. Brunner and P. Howard. Russell's alternative to the axiom of choice. *Z. Math. Logik Grundl. Math.*, 38:529–534, 1992.

[Bus93] G. Buskes. The Hahn–Banach Theorem surveyed. *Dissertationes Math.* 327, 1993.

[Cha72] R.E. Chandler. An alternative construction of βX and νX. *Proc. Amer. Math. Soc.* 32:315–318, 1972.

[Chu27] A. Church. Alternatives to Zermelo's Assumption. *Trans. Amer. Math. Soc.*, 29:178–208, 1927.

[Coh63/64] P. J. Cohen. The independence of the continuum hypothesis, I, II. *Proc. Nat. Acad. Sci. USA*, 50:1143–1148, 1963, 51:105–110, 1964.

[Coh66] P. J. Cohen. *Set Theory and the Continuum Hypothesis.* New York, 1966.

[Coh2002] P. Cohen. The discovery of forcing. *Rocky Mountain J. Math.*, 32:1071–1100, 2002.

[Com68] W.W. Comfort. A theorem of Stone–Čech and a theorem of Tychonoff, without the axiom of choice; and their realcompact analogues. *Fund. Math.*, 68:97–110, 1968.

[Ded1888] R. Dedekind. Was sind und was sollen die Zahlen? Vieweg Verlag, 1888.

[DCr2002] O. De la Cruz. Finiteness and choice. *Fund. Math.*, 173:57–76, 2002.

[DHHRS2002] O. De la Cruz, E. Hall, P. Howard, J. E. Rubin and A. Stanley. Definitions of compactness and the axiom of choice. *J. Symb. Logic*, 67:143–161, 2002.

[DHHKR2002] O. De la Cruz, E. Hall, P. Howard, K. Keremedis and J.E. Rubin. Products of compact spaces and the axiom of choice. *Math. Logic Quart.*, 48:508–516, 2002.

[DHHKR2003] O. De la Cruz, E. Hall, P. Howard, K. Keremedis and J.E. Rubin. Metric spaces and the axiom of choice. *Math. Logic Quart.*, 49:455–466, 2003.

[DHHKR2003a] O. De la Cruz, E. Hall, P. Howard, K. Keremedis and J.E. Rubin. Products of compact spaces and the axiom of choice II. *Math. Log. Quart.*, 49:57–71, 2003.

[DekGr56] T.J. Dekker and J. de Groot. Decompositions of a sphere. *Fund. Math.*, 43:185–194, 1956.

[Dei2005] O. Deiser. Der Multiplikationssatz der Mengenlehre. *Jahresber. Deutsche Math.-Vereinigung*, 107:88–109, 2005.

[Den2003] J.T. Denniston. AFT implies the Axiom of Choice for Classes. Manuscript Oct. 2003.

[DoMo99] J. Dodu and M. Morillon. The Hahn–Banach property and the axiom of choice. *Math. Logic Quart.*, 45:299–314, 1999.

[Eng89] R. Engelking. General Topology. *Heldermann Verlag Berlin*, 1989.

[Erd40] P. Erdös. The dimension of the rational points in Hilbert space. *Ann. Math.*, 41:734–736, 1940.

[Erd80] P. Erdös. Some combinatorial problems in geometry. *Springer Lect. Notes Math.*, 792:46–53, 1980.

[Ern97] M. Erné. Prime ideal theorems and systems of finite character. *Comment. Math. Univ. Carolinae*, 38:513–536, 1997.

[Ern2001] M. Erné. Constructive order theory. *Math. Logic Quart.*, 47:211–222, 2001.

[Ern200?] M. Erné. Prime and maximal ideals of partially ordered sets. Preprint 2005.

[Fal81] K.J. Falconer. The realization of distances in measurable subsets covering $\mathbb{R}^n$. *J. Comb. Theory (A)*, 31:184–189, 1981.

[Fef2000] S. Fefermann. Mathematical Intuition vs. Mathematical Monsters. *Synthese*, 125:317–332, 2000.

[Fel71] U. Felgner. *Models of ZF–set theory*, Springer Lecture Notes 233, 1971.

[Fel79] U. Felgner. Mengelehre. Wiss. Buchges. Darmstadt, 1979.

[FeJe73] U. Felgner and T.J. Jech. Variants of the axiom of choice in set theory with atoms. *Fund. Math.*, 79:79–85, 1973.

[FeSch84] U. Felgner and K. Schulz. Algebraische Konsequenzen des Determiniertheit–Axioms. *Arch. Math.*, 42:557–563, 1984.

[FeTr99] U. Felgner and J.K. Truss. The independence of the prime ideal theorem from the order–extension principle. *J. Symbolic Logic*, 64:199–215, 1999.

[FoMo98] J. Fossy and M. Morillon. The Baire category property and some notions of compactness. *J. London Math. Soc.*, 57:1–19, 1998.

[FoWe91] M. Foreman and F. Wehrung. The Hahn–Banach theorem implies the existence of non–Lebesgue measurable sets. *Fund. Math.*, 138:13–19, 1991.

[Fra37] A. Fraenkel. Ueber eine abgeschwaechte Fassung des Auswahlaxioms. *J. Symb. Logic*, 2:1–25, 1937.

[FrBaLe73] A.A. Fraenkel, Y. Bar–Hillel and A. Levy. *Foundations of Set Theory*. 2nd ed., North–Holland, 1973.

[GaKo91] F. Galvin and P. Komjáth. Graph colorings and the axiom of choice. *Period. Math. Hungar.*, 22:71–75, 1991.

[GiHe2004] E. Giuli and H. Herrlich. On closure operators, the reals, and choice. *Quaest. Math.*, 27:3–3, 2004.

[Goed38] K. Gödel. The consistency of the axiom of choice and of the generalized continuum–hypothesis. *Proc. Nat. Acad. Sci. USA*, 24:556–557, 1938.

[Goed39] K. Gödel. Consistency–proof for the generalized continuum hypothesis. *Proc. Nat. Acad. Sci. USA*, 25:220–224, 1939.

[Goed47] K. Gödel. What is Cantor's Continuum Problem? *Amer. Math. Monthly*, 54:515–525, 1947.

[Gol85] R. Goldblatt. On the Role of the Baire Category Theorem and Dependent Choice in the Foundations of Logic. *J. Symbolic Logic*, 50:412–422.

[GoMy78] N. Goodman and J. Myhill. Choice implies excluded middle. *Zeitschr. math. Logik Grundlagen Math.*, 24:461, 1978.

[Goo68] R. L. Goodstein. Existence in mathematics. In *Logic and Foundations of Mathematics*, pages 70–82. Wolters–Noordhoff Publ., Groningen, 1968.

[GoTr95] C. Good and I.J. Tree. Continuing horrors of topology without choice. *Topol. Appl.*, 63:79–90, 1995.

[GoTrWa98] C. Good, I.J. Tree and W.S. Watson. On Stone's Theorem and the Axiom of Choice. *Proc. Amer. Math. Soc.*, 126:1211–1218, 1998.

[Grae86] G. Grätzer. Birkhoff's representation theorem is equivalent to the axiom of choice. *Algebra Univers.*, 23:58–60, 1986.

[Gut2003] G. Gutierres. Sequential topological conditions in $\mathbb{R}$ in the absence of the axiom of choice. *Math. Log. Quart.*, 49:293–298, 2003.

[Gut2004] G. Gutierres. O Axioma da Escola Numerável em Topologia. Thesis, Univ. Coimbra, 2004.

[Gut2004a] G. Gutierres. On first and second countable spaces and the axiom of choice. *Topology Appl.*, 143:93–103, 2004.

[Haeu83] A. Häussler. Defining cardinal addition by $\leq$–formulas. *Fund. Math.*, 115:195–205, 1983.

[HaMo90] L. Haddad and M. Morillon. L'axiome de normalité pour les espaces totalement ordonnés. *J. Symbolic Logic*, 55:277–283, 1990.

[Hal64] J.D. Halpern. The independence of the axiom of choice from the Boolean prime ideal theorem. *Fund. Math.*, 66:57–66, 1964.

[Hal66] J.D. Halpern. Bases for vector spaces and the axiom of choice. *Proc. Amer. Math. Soc.*, pages 670 - 673, 1966.

[HaHo70] J.D. Halpern and P. Howard. Cardinals m such that $2m = m$. *Proc. Amer. Math. Soc.*, 26:487–490, 1970.

[HaLe71] J.D. Halpern and A. Lévy. The Boolean prime ideal theorem does not imply the axiom of choice. *Symp. Pure Math.*, 1:83–134, 1971.

[Halp51] I. Halperin. Non–measurable sets and the equation $f(x+y) = f(x)+f(y)$. *Proc. Amer. Math. Soc.*, 2:221–224, 1951.

[Ham05] G. Hamel. Eine Basis aller Zahlen und die unstetigen Lösungen der Funktionalgleichung: $f(x+y) = f(x) + f(y)$. *Math. Annalen*, 60:459–462, 1905.

[Hard06] G.H. Hardy. The continuum and the second number class. *Proc. London Math. Soc. Series 2*, 4:10–17, 1906.

[Har15] F. Hartogs. Über das Problem der Wohlordnung. *Mathem. Annalen*, 76:438–443, 1915.

[Hau14] F. Hausdorff. Grundzüge der Mengenlehre. Berlin 1914.

[Her68] H. Herrlich. Topologische Reflexionen und Coreflexionen. *Springer Lecture Notes Math.* 78, 1968.

[Her83] H. Herrlich. Are there convenient subcategories of TOP? *Topol. Appl.*, 15:263–271, 1983.

[Her96] H. Herrlich. Compactness and the axiom of choice. *Applied Categorical Structures*, 4:1–14, 1996.

[Her96a] H. Herrlich. An effective construction of a free z ultrafilter. Papers on Gen. Topology and Appl. 11th Summer Conf. Univ. Southern Maine. (Eds. S. Andima, R.C. Flagg, G. Itzkovitz, Y. Kong, R. Kopperman, and P. Misra). *Annals New York Acad. Sci*, 806:201–206, 1996.

[Her97] H. Herrlich. The Ascoli Theorem is equivalent to the Boolean Prime Ideal Theorem. *Rostock Math. Kolloq.*, 51:137–140, 1997.

[Her97a] H. Herrlich. The Ascoli Theorem is equivalent to the Axiom of Choice. *Seminarberichte Fb. Math. Univ. Hagen*, 62:97–100, 1997.

[Her97b] H. Herrlich. Choice principles in elementary topology and analysis. *Comment. Math. Univ. Carolinae*, 38:545–552, 1997.

[Her2002] H. Herrlich. Products of Lindelöf T_2-spaces are Lindelöf – in some models of ZF. *Comment. Math. Univ. Carolinae*, 43:319–333, 2002.

[Her2003] H. Herrlich. The axiom of choice holds iff maximal closed filters exist. *Math. Logic Quart.* 3:323–324, 2003.

[Her2005] H. Herrlich. Zur Existenz maximaler Filter und Ideale. *Seminarberichte Fb. Math. Univ. Hagen*, 76:31–42, 2005.

[HeKe99] H. Herrlich and K. Keremedis. Products, the Baire category, and the axiom of dependent choice. *Comment. Math. Univ. Carolinae*, 40:771–775, 1999.

[HeKe99a] H. Herrlich and K. Keremedis. Powers of **2**. *Notre Dame Journal of Formal Logic*, 40:346–351, 1999.

[HeKe2000] H. Herrlich and K. Keremedis. On countable products of finite Hausdorff spaces. *Math. Logic Quart.*, 46:537–542, 2000.

[HeKe2000a] H. Herrlich and K. Keremedis. The Baire category theorem and choice. *Topology Appl.*, 108:157–167, 2000.

[HeKeTa2002] H. Herrlich, K. Keremedis and E. Tachtsis. Striking differences between ZF and ZF + weak choice in view of metric spaces. *Quaest. Math*, 25:405–420, 2002.

[HeKeTa2005] H. Herrlich, K. Keremedis and E. Tachtsis. Countable sums and products of Loeb and selective metric spaces. *Comment. Math. Univ. Carolinae*, 46:373–384, 2005.

[HeRh2005] H. Herrlich and Y.T. Rhineghost. Graph–Coloring and Choice. A Note on a Note by Shelah and Soifer. *Quaest. Math*, 28:317–319, 2005.

[HeSt97] H. Herrlich and G.E. Strecker. When is $\mathbb{N}$ Lindelöf ? *Comment. Math. Univ. Carolinae*, 38:553–556, 1997.

[HeSt97a] H. Herrlich and G.E. Strecker. Categorical Topology — its Origins, as examplified by the unfolding of the Theory of Topological Reflections and Coreflections before 1971. In: Handbook of the History of General Topology. Vol 1. (Eds. C.E. Aull and R. Lowen). Kluwer Acad. Publ. 1997, 255–341.

[HeStr97] H. Herrlich and J. Steprāns. Maximal filters, continuity and choice principles. *Quaest. Math.*, 20:697–705, 1997.

[HewSt69] E. Hewitt and K. Stromberg. Real and Abstract Analysis. Springer 1969.

[Hey56] A. Heyting. Intuitionism, an introduction. In *Studies in Logic and the Foundation of Mathematics.* North Holland Publ. Comp., Amsterdam, 1956.

[Hic76] J.L. Hickman. The construction of groups in models of set theory that fail the axiom of choice. *Bull. Austral. Math. Soc.*, 14:199–232, 1976.

[Hic78] J.L. Hickman. Dedekind-finite fields. *Bull. Austr. Math. Soc.*, 19:117–124, 1978.

[Hil1900] D. Hilbert. Mathematische Probleme. Vortrag: Internat. Math.-Kongr. Paris 1900 *Nachr. Akad. Wiss. Göttingen*, 253–297. English translation: *Bull. Amer. Math. Soc.*, 2(8):437–479, 1902.

[Hil26] D. Hilbert. Über das Unendliche. *Math. Ann.*, 95:161–190, 1926.

[How90] P. Howard. Definitions of compact. *J. Symb. Logic*, 55:645–655, 1990.

[How92] P. Howard. The axiom of choice for countable collections of countable sets does not imply the countable union theorem. *Notre Dame J. Formal Logic*, 33:236–243, 1992.

[HoRu96] P. Howard and J.E. Rubin. The Boolean prime ideal theorem plus countable choice do not imply dependent choice. *Math. Logic Quart.*, 42:410–420, 1996.

[HoRu98] P. Howard and J.E. Rubin. Consequences of the Axiom of Choice. *Amer. Math. Soc.* 1998. Project Homepage.
`http://www.math.purdue.edu/~jer/cgi-bin/conseq.html` or
`http://www.dragon.emich.edu/~phoward/conseq.html`.

[HoYo89] P. Howard and M. Yorke. Definitions of finite. *Fund. Math.*, 133:169–177, 1989.

[HoeHo94] H. Höft and P. Howard. Well ordered subsets of linearly ordered sets. *Notre Dame J. Formal Logic*, 35:413–425, 1994.

[HKRR98] P. Howard, K. Keremedis, H. Rubin and J.E. Rubin. Versions of normality and some weak forms of the axiom of choice. *Math. Logic Quart.*, 44:367–382, 1998.

[HKRR98a] P. Howard, K. Keremedis, H. Rubin and J.E. Rubin. Disjoint unions of topological spaces and choice. *Math. Logic Quart.*, 44:493–598, 1998.

[HKRST2001] P. Howard, K. Keremedis, J.E. Rubin, A. Stanley and E. Tachtsis. Non–constructive properties of the real numbers. *Math. Logic Quart.*, 47:423–431, 2001.

[Jae65] M. Jaegermann. The axiom of choice and two definitions of continuity. *Bull. l'Acad. Polon. Sci. Ser. Sci., Math., Astr., Phys.*, 13:699–704, 1965.

[Jec68] T.J. Jech. Eine Bemerkung zum Auswahlaxiom. *Časopis pro pěst. matem.*, 93:30–31, 1968.

[Jec73] T.J. Jech. *The Axiom of Choice.* North Holland Studies in Logic and the Foundations of Math., Amsterdam, 1973.

[Jen98] R.B. Jensen. Exploring the Infinite: Developments in Set Theory. *DMV-Mitteilungen*, 2:52–55, 1998.

[Jon42] F.B. Jones. Connected and disconnected plane sets and the functional equation $f(x) + f(y) = f(x + y)$. *Bull. Amer. Math. Soc.*, 48:115–120, 1942.

[Jur65] W. Jurkat. On Cauchy's functional equation. *Proc. Amer. Math. Soc.*, 16:683–686, 1965.

[Kac36/37] M. Kac. Une remarque sur les équations fonctionelles. *Comment. Math. Helvet.*, 9:170–171, 1936/37.

[Kec84] A.S. Kechris. The axiom of deteminancy implies dependent choices in $L(\mathbb{R})$. *J. Symb. Logic*, 49:161–173, 1984.

[Kel50] J.L. Kelley. Tychonoff's theorem implies **AC**. *Fund. Math.* 37:75–76, 1950.

[Ker96] K. Keremedis. Bases for vector spaces over the two-element field and the axiom of choice. *Proc. Amer. Math. Soc.*, 124:2527–2531, 1996.

[Ker98] K. Keremedis. Extending independent sets to bases and the axiom of choice. *Math. Logic Quart.*, 44:92–98, 1998.

[Ker2000] K. Keremedis. The compactness of $2^{\mathbb{R}}$ and the axiom of choice. *Math. Logic Quart.*, 46:569–571, 2000.

[Ker2000a] K. Keremedis. On Weierstrass compact pseudometric spaces and a weak form of the axiom of choice. *Topology Appl.*, 108:75–78, 2000.

[Ker200?] K. Keremedis. Countable disjoint unions in topology and weak forms of the axiom of choice. *Archiv Math. Logic.*

[Ker2003] K. Keremedis. Some weak forms of the Baire category theorem. *Math. Logic Quart.*, 49:369–374, 2003.

[Ker2005] K. Keremedis. Tychonoff products of the two element set and some weakenings of the Boolean prime ideal theorem. *Manuscript*, June 2005.

[KeTa99] K. Keremedis and E. Tachtsis. On the extendibility of closed filters in T_1-spaces and the existence of well orderable filter bases. *Comment. Math. Univ. Carolinae*, 40:343–353, 1999.

[KeTa2001] K. Keremedis and E. Tachtsis. Some weak forms of the axiom of choice restricted to the real line. *Math. Log. Quart.*. 47:413–422, 2001.

[KeTa2001a] K. Keremedis and E. Tachtsis. Compact metric spaces and weak forms of the axiom of choice. *Math. Logic Quart.*, 47:117–128, 2001.

[KeTa2003] K. Keremedis and E. Tachtsis. Choice principles for special subsets of the real line. *Math. Logic Quart.*, 49:444–454, 2003.

[KeTa2004] K. Keremedis and E. Tachtsis. Topology in the absence of the axiom of choice. *Mathematica Japonica*, 59:357–406, 2004.

[KeTa2005] K. Keremedis and E. Tachtsis. Countable sums and products of metrizable spaces in ZF. *Math. Logic Quart.* 51:95–103, 2005.

[Kes51] H. Kestelman. Automorhpisms of the field of complex numbers. *Proc. London Math. Soc. 2*, 53:1–12, 1951.

[KiWa55] W. Kinna and K. Wagner. Über eine Abschwächung des Auswahlpostulates. *Fund. Math.*, 42:75–82, 1955.

[Kle77] E.M. Kleinberg. Infinitary combinatorics and the axiom of determinateness. *Lecture Notes in Mathematics*, 612, 1977.

[Kli58] G. Klimowsky. El Theorema de Zorn y la existencia de filtros e ideales maximales en los reticulados distributivos. *Rev. Union Mat. Argentina*, 18:160–164, 1958.

[Kro81] M. Krom. Equivalents of a Weak Axiom of Choice. *Notre Dame J. Formal Logic*, 22:283–285, 1981.

[Kro86] M. Krom. A linearly ordered topological space that is not normal. Notre Dame J. Formal Logic 27:12–13, 1986.

[Laeu62/63] H. Läuchli. Auswahlaxiom in der Algebra. *Commentarii Math. Helvetici*, 37:1–18, 1962/1963.

[Laeu71] H. Läuchli. Coloring infinite graphs and the Boolean Prime Ideal Theorem. *Israel J. Math.*, 9:422–429, 1971.

[Lev58] A. Levy. The independence of various definitions of finiteness. *Fund. Math.*, 46:1–13, 1958.

[Lit26] J.E. Littlewood. *The elements of the theory of real functions, being notes of lectures delivered in the University of Cambridge.* Heffer, Cambridge, 1926.

[Loe65] P.A. Loeb. A new proof of the Tychonoff theorem. *Amer. Math. Monthly*, 72:711–717, 1965.

[LoRy51] J. Łoś and C. Ryll–Nardzewski. On the application of Tychnonoff's Theorem in mathematical proofs. *Fund. Math.*, 38:233–237, 1951.

[LoRy55] J. Łoś and C. Ryll–Nardzewski. Effectiveness of the representation theory for Boolean algebras. *Fund. Math.*, 41:49–56, 1955.

[Lux69] W.A.J. Luxemburg. Reduced powers of the real number system and equivalents of the Hahn–Banach theorem. *Applications of Model Theory to Algebra, Analysis and Probability.* (Ed. W.A. Luxemburg), 123–137, 1969.

[Mad88] P. Maddy. Believing the axioms I. *J. Symb. Logic*, 53:481–511, 1988.

[Mad88a] P. Maddy. Believing the axioms II. *J. Symb. Logic* 53:736–764, 1988.

[MaMy2001] V.W. Marek and J. Mycielski. Foundations of mathematics in the twentieth century. *Amer. Math. Monthly*, 108:449–468, 2001.

[Mar2003] D.A. Martin. A simple proof that determinancy implies Lebesgue measurability. *Rend. Sem. Mat. Univ. Politec., Torino.* 61:393–397, 2003.

[Mig81] R. Mignone. Ultrafilters resulting from the axiom of determinateness. *Proc. London Math. Soc. (3)*, 43:582–605, 1981.

[Mon75] G.P. Monro. Independence results concerning Dedekind–finite sets. *J. Austral. Math. Soc. (Series A)*, 19:35–46, 1975.

[Moo82] G.H. Moore. Zermelo's Axiom of Choice, its Origins, Development, and Influence. *Springer Studies in the History of Mathem. and Physic. Sci.* 8, 1982.

[Mos39] A. Mostowski. Über die Unabhängigkeit des Wohlordnungssatzes vom Ordnungsprinzip. *Fund. Math.*, 32:201–252, 1939.

[Mos45] A. Mostowski. Axiom of choice for finite sets. *Fund. Math.*, 33:137–168, 1945.

[Mos67] A. Mostowski. Recent Results in Set Theory. In *Problems in the Philosophy of Mathematics*, (Ed. I. Lakatos), Proc. Intern. Coll. Phil. Sci., pages 82–108, London, 1967. North–Holland Publ. Co.

[Myc61] J. Mycielski. Some remarks and problems on the coloring of infinite graphs and the theorem of Kuratowski. *Acta Math. Acad. Sci. Hungary* 12:125–129, 1961.

[Myc64] J. Mycielski. On the axiom of determinateness. *Fund. Math.*, 53:205–224, 1964.

[Myc64a] J. Mycielski. Two remarks on Tychonoff's product theorem. *Bull. Acad. Polon. Sci. Sér. Sci. Math. Ast. Phys.* 12:439–441, 1964.

[Myc66] J. Mycielski. On the axiom of determinateness II. *Fund. Math.*, 59:203–212, 1966.

[MySt62] J. Mycielski and H. Steinhaus. A mathematical axiom contradicting the Axiom of Choice. *Bull. Acad. Polon. Sci. Ser. Sci. Mat. Astr. Phys.* 10:1–3, 1962.

[MySw64] J. Mycielski and S. Świerczkowski. On the Lebesgue measurability and the axiom of determinateness. *Fund. Math.*, 54:67–71, 1964.

[NaBe97] L. Narici and E. Beckenstein. The Hahn–Banach theorem: its life and times. *Topology Appl.*, 77:193–211, 1997.

[Oxt61] J.C. Oxtoby. Cartesian products of Baire spaces. *Fund. Math.*, 49:157–166, 1961.

[Oxt80] J.C. Oxtoby. *Measure and Category.* Second Edition. Springer Graduate Texts in Math. 2, 1980.

[Paw91] J. Pawlikowski. The Hahn–Banach Theorem implies the Banach–Tarski Paradox. *Fund. Math.*, 138:21–22, 1991.

[Pin72] D. Pincus. Independence of the Prime Ideal Theorem from the Hahn Banach Theorem. *Bull. Amer. Math. Soc.*, 78:766–770, 1972.

[Pin77] D. Pincus. Adding Dependent Choice. *Ann. Math. Logic*, 11:105, 1977.

[Pot90] M.D. Potter. *Sets.* Clarendon Press, 1990.

[Qui48] W.V.O. Quine. On what there ist. *Review of Metaphysics*, 2:21–38, 1948.

[Rai84] J. Raisonnier. A mathematical proof of Shelah's theorem on the measure problem and related results. *Israel J. Math.*, 48:48–56, 1984.

[Rhi2001] Y.T. Rhineghost. The naturals are Lindelöf iff Ascoli holds. In: Categorical Perspectives. (Eds. J. Koslowski and A. Melton). Birkhäuser 2001, 191–196.

[Rhi2002] Y.T. Rhineghost. The Boolean Prime Ideal Theorem holds iff maximal open filters exist. *Cah. Topol. Géom. Différ. Catég.*, 43:313–315, 2002.

[Rob47] R.M. Robinson. On the decomposition of spheres. Fund. Math., 34:246–260, 1947.

[Rub60] H. Rubin. Two propositions equivalent to the axiom of choice only under both the axiom of extensionality and regularity. *Notices Amer. Math. Soc.*, 7:380, 1960.

[RuRu85] H. Rubin and J.E. Rubin. Equivalents of the axiom of choice, II. *North Holland Studies in Logic and Foundation of Math.* 116, 1985.

[RuSc54] H. Rubin and D. Scott. Some topological theorems equivalent to the Boolean Prime Ideal Theorem. *Bull. Amer. Math. Soc.*, 60:389, 1954.

[Rus07] B. Russell. On some difficulties in the theory of transfinite numbers and order types. *Proc. London Math. Soc.*, 4:29–53, 1907.

[Rus11] B. Russell. Sur les axiomes de l'infini et du transfini. *Soc. math. France, Comptes rendues des séances*, 2:22–35, 1911.

[Sal74] S. Salbany. On compact* spaces and compactifications. *Proc. Amer. Math. Soc.*, 45:274–280, 1974.

[SaZi2002] M. Sardella and G. Ziliotti. What's the price of a nonmeasurable set? *J. Math. Bohem.*, 127:41–48, 2002.

[Saw82] W.W. Sawyer. Prelude to Mathematics. Dover 1982.

[Sch92] E. Schechter. Two topological equivalents of the axiom of choice. *Z. Math. Logik Grundl. Math.*, 38:555–557, 1992.

[Sch97] E. Schechter. *Handbook of Analysis and its Foundation.* Acad. Press, 1997.

[Sco54] D. Scott. The theorem on maximal ideals in lattices and the axiom of choice. *Bull. Amer. Math. Soc.*, 60:83, 1954.

[She85] S. Shelah. On measure and category. *Israel J. Math.*, 52:110–114, 1985.

[ShSo2003] S. Shelah and A. Soifer. Axiom of choice and chromatic number of the plane. *J. Comb. Theory, Ser. A*, 103:387-391, 2003.

[Sie16] W. Sierpiński. Sur le rôle de l'axiome de M. Zermelo dans l'Analyse moderne. Compt. Rend. Hebd. Sean. l'Acad. Sci., Paris, 193:688–691, 1916.

[Sie18] W. Sierpiński. L'axiome de M. Zermelo et son rôle dans la Théorie des Ensembles et l'Analyse. Bulletin de l'Academie des Sciences de Cracovie, Cl. Sci. Math., Sér. A, 97–152, 1918.

[Sie20] W. Sierpiński. Sur l'équation fonctionelle $f(x+y) = f(x) + f(y)$. *Fund. Math.*, 1:116–122, 1920.

[Sie21] W. Sierpiński. Les exemples effectives et l'axiome de choix. *Fund. Math.*, 2:112–118, 1921.

[Sie27] V. Sierpiński. Sur un problème conduisant á un ensemble non mesurable. *Fund. Math.*, 10:177–179, 1927.

[Sie29] V. Sierpiński. Sur un paradoxe de M.J. von Neumann. *Fund. Math.*, 35:203–207, 1948.

[Sie30] V. Sierpiński. Sur une propriété de la décomposition de M. Vitali. *Mathematica*, 3:30–32, 1930.

[Sie38] V. Sierpiński. Fonctions additives non complètement additives et fonctions non mesurables. *Fund. Math.*, 30:96–99, 1938.

[Sie45] V. Sierpiński. Sur le paradox de MM. Banach et Tarski. *Fund. Math.*, 33:229–234, 1945.

[Sie47] W. Sierpiński. L'hypothèse généralisée du continu et l'axiome du choix. *Fund. Math.*, 34:1–5, 1947.

[Sie47a] W. Sierpiński. Sur une proposition qui entraine l'existence des ensembles non–mesurables. *Fund. Math.*, 34:157–162, 1947.

[Sie48] W. Sierpiński. Sur l'équivalence des ensembles par décomposition en deux parties. *Fund. Math.*, 35:151–158, 1948.

[Sie58] W. Sierpiński. *Cardinal and Ordinal Numbers.* Polska Akad. Nauk Monogr. Matem. 34, Warszawa, 1958.

[Sob61] B. Sobociński. A theorem on Hartog's Alephs. *Notre Dame J. Formal Logic*,2:255–258, 1961.

[Soi2003] A. Soifer. Chromatic number of the plane: Its past and future. *Congr. Numerantium*, 160:69–82, 2003.

[Sol65] R.M. Solovay. The measure problem. *Notices Amer. Math. Soc.*, 12:217, 1965.

[Sol70] R.M. Solovay. A model of set–theory in which every set of reals is Lebesgue measurable. *Annals Math.*, 92:1–56, 1970.

[Spe54] E. Specker. Verallgemeinerte Kontinuumshypothese und Auswahlaxiom. *Archiv Math.*, 5:332–337, 1954.

[Spe57] E. Specker. Zur Axiomatik der Mengenlehre (Fundierungs- und Auswahlaxiom). *Zeitschr. f. math. Logik und Grundl. der Math.*, 3:173–210, 1957.

[Str79] K. Stromberg. The Banach–Tarski Paradox. *Amer. Math. Monthly*, 86:151–161, 1979.

[Tach2002] E. Tachtsis. Disasters in metric topology without choice. *Comment. Math. Univ. Carolinae*, 43:165–174, 2002.

[Tar24] A. Tarski. Sur quelques théorèmes qui équivalent à l'axiome du choix. *Fund. Math.*, 5:147–154, 1924.

[Tar24a] A. Tarski. Sur les ensembles finis. *Fund. Math.*, 6:45–95, 1924.

[Tar38] A. Tarski. Ein Überdeckungssatz für endliche Mengen nebst einigen Bemerkungen für Definitionen der Endlichkeit. *Fund. Math.*, 30:156–163, 1938.

[Tar38a] A. Tarski. Über das absolute Maß linearer Punktmengen. *Fund. Math.*, 30:218–234, 1938.

[Tar39] A. Tarski. On well–ordered subsets of any set. *Fund. Math.*, 32:176–183, 1939.

[Tar54] A. Tarski. Theorems on the existence of successors of cardinals, and the Axiom of Choice. *Indag. Math.*, 16:26–32, 1954.

[Tar65] A. Tarski. On the existence of large sets of Dedekind cardinals. *Notices Amer. Math. Soc.*, 12:719, 1965.

[Tru74] J.K. Truss. Classes of Dedekind finite cardinals. *Fund. Math.*, 84:187–208, 1974.

[Tru95] J.K. Truss. The structure of amorphous sets. *Ann. Pure Appl. Logic*,73:191–233, 1995.

[vDou85] E. K. van Douwen. Horrors of topology without AC: A nonnormal orderable space. *Proc. AMS*, 95:101–105, 1985.

[vNeu25] J. von Neumann. Eine Axiomatisierung der Mengenlehre. *J. Math.*, 154:219–240, 1925.

[vNeu29] J. von Neumann. Zur allgemeinen Theorie des Maßes. *Fund. Math.*, 30:73–116, 1929.

[Ver94] M. Vervoort. An elementary construction of an ultrafilter on $\aleph_1$ using the Axiom of Determinateness.
`http://staff.science.uva.nl/~vervoort/ultrafilter.pdf`.

[Vit05] G. Vitali. *Sul Problema della misure dei gruppi di punti di una retta.* Bologna, 1905.

[Wag86] S. Wagon. The Banach–Tarski Paradox. *Cambr. Univ. Press. Encyl. Mathem. and its Appl.* 24, 1986.

[War62] L.E. Ward. A weak Tychonoff Theorem and the Axiom of Choice. *Proc. Amer. Math. Soc.*, 13:757–758, 1962.

[Wila70] A. Wilansky. *Topology for Analysts.* Ginn and Co., 1970.

[Wil70] S. Willard. *General Topology.* Addison–Wesley Publ. Co., 1970.

[Zer04] E. Zermelo. Beweis, daß jede Menge wohlgeordnet werden kann. *Math. Annalen*, 59:514–516, 1904.

[Zer08] E. Zermelo. Neuer Beweis der Möglichkeit einer Wohlordnung. *Math. Annalen*, 65:107–128, 1908.

[Zer08a] E. Zermelo. Untersuchungen über die Grundlagen der Mengenlehre: I. *Math. Annalen* 65:261–281, 1908.

Selected Books and Longer Articles

Felgner, Ulrich
1971 *Models of ZF Set Theory.*
Springer Lectures Notes Math., **233**

Howard, Paul, and Jean E. Rubin
1998 *Consequences of the Axiom of Choice.*
Amer. Math. Soc. Math. Surveys and Monogr. **59**.

Jech, Thomas J.
1973 *The Axiom of Choice.*
North Holland Studies in Logic and the Foundations of Math. **75**.

Kanovej, V.G.
1984 The Axiom of Choice and the Axiom of Determinateness (Russian)
Zbl. 0599.03053; MR0767 261 (86h:03085).

Keremedis, Kyriakos and Eleftherios Tachtsis
2004 *Topology in the absence of the Axiom of Choice.*
Mathematica Japonica Scientia, **59**, 357–406.

Moore, Gregory H.
1982 *Zermelo's Axiom of Choice, its Origins, Development, and Influence.*
Springer Studies in the History of Mathem. and Physics. Sci. **8**.

Oxtoby, John C.
1980 *Measure and Category.*
Second Edition. Springer Graduate Texts in Math.

Rubin, Herman, and Jean E. Rubin
1985 *Equivalence of the Axiom of Choice, II.*
North Holland Studies in Logic and the Found. of Math. **116**.

Schechter, Eric
1997 *Handbook of Analysis and its Foundations.*
Acad. Press.

Sierpiński, Waclaw

1918 *L'axiome de M. Zermelo et son rôle dans la Théorie des Ensembles et l'analyse.* Bull. Acad. Sci. Cracovie, 97–152.

1958 *Cardinal and Ordinal Numbers.*
Polska Akad. Nauk. Monogr. Mathem. **34**.

Wagon, Stan

1986 *The Banach–Tarski Paradox.*
Cambr. Univ. Press. Encycl. Mathem. and its Appl. **24**.

List of Symbols

$|X|$ cardinality of the set X

Ordering of cardinals:

$|X| = |Y| \Leftrightarrow \exists f\colon X \to Y$ bijective

$|X| \leq |Y| \Leftrightarrow \exists f\colon X \to Y$ injective

$|X| < |Y| \Leftrightarrow |X| \leq |Y|$ and $|X| \neq |Y|$

$|X| \leq^* |Y| \Leftrightarrow X = \emptyset$ or $\exists f\colon Y \to X$ surjective

etc.

$\aleph$ Aleph (cardinal of a well–orderable infinite set)

$\aleph_0 = |\mathbb{N}|$

Ord = Class of all ordinals

$\alpha = \{\beta \in \mathrm{Ord} \mid \beta < \alpha\}$ for ordinals α

$2 = \{0, 1\}$

2 is the discrete topological space (or, sometimes, the lattice) with underlying set 2.

Special sets:

$\mathbb{N}$ = set of natural numbers $0, 1, 2, ...$

$\mathbb{N}^+ = \mathbb{N} \backslash \{0\}$

$\mathbb{Z}$ = set of integers

$\mathbb{Q}$ = set of rational numbers

$\mathbb{R}$ = set of real numbers

$\mathbb{C}$ = set of complex numbers

$[a, b] = \{x \in \mathbb{R} \mid a \leq x \leq b\}$

$[a, b) = \{x \in \mathbb{R} \mid a \leq x < b\}$

$(a, b) = \{x \in \mathbb{R} \mid a < x < b\}$

$\mathcal{P}X$	$= \{A \mid A \subseteq X\}$	powerset of X
$\mathcal{P}_0 X$	$= \mathcal{P}X \setminus \{\emptyset\}$	
$\mathcal{P}_{\text{fin}} X$	$= \{A \in \mathcal{P}X \mid A \text{ finite}\}$	
$X \cap Y$	$= \{z \mid z \in X \text{ and } z \in Y\}$	intersection
$X \cup Y$	$= \{z \mid z \in X \text{ or } z \in Y\}$	union
$X \uplus Y$	$= (X \times \{0\}) \cup (Y \times \{1\})$	disjoint union
$X \setminus Y$	$= \{z \mid z \in X \text{ and } z \notin Y\}$	different
$X \Delta Y$	$= (X \setminus Y) \cup (Y \setminus X)$	symmetric difference
$\bigcup_{i \in I} X_i$	$= \{z \mid \exists i \in I \ \ z \in X_i\}$	union
$\biguplus_{i \in I} X_i$	$= \bigcup_{i \in I} (X_i \times \{i\})$	disjoint union, sum
$\prod_{i \in I} X_i$	$= \{f \colon I \to \bigcup_{i \in I} X_i \mid \forall i \in I \ \ f(i) \in X_i\}$	product
$[0,1]^I$	= Hilbert cubes	
$\mathbf{2}^I$	= Cantor cubes	
$C(X,Y)$	$= \{f \colon X \to Y \mid f \text{ continuous}\}$	
$C_{co}(X,Y)$	$= (C(X,Y), \tau_{co})$	
$C(X)$	$= C(X, \mathbb{R})$	
$C^*(X)$	$= \{f \in C(X) \mid f \text{ bounded}\}$	
τ_{co}	compact open topology on $C(X,Y)$	
$A \approx B$	A congruent with B	
$A \sim_e B$	A equidecomposable with B	

List of Axioms

In brackets the corresponding *form numbers* in [HoRu98]. Diagrams showing implications between some of the axioms can be found in 2.21, 3.4, 4.58, 5.10, 5.25, and in A1 – A6.

AC	*Axiom of Choice*	1.1	[1]
AC($c\mathbb{R}$)		4.55	
AC(fin)		2.6	[62]
AC(n)		2.6	[45]
AC($\mathbb{R}$)		E 1.Sec.1.1	[79]
AC(X)		E 1.Sec.1.1	
AD	*Axiom of Determinateness*	7.12	
AH	*Aleph–Hypothesis*	2.19	
AH(0)	*Special Aleph–Hypothesis*	2.19	
AMC	*Axiom of Multiple Choice*	2.4 and 2.7	[67]
BTP	*Banach–Tarski Paradox*	5.23	[309]
BP	*Baire Property for subsets of* $\mathbb{R}$,	Preface (footnote 15)	[-142]
CC	*Axiom of Countable Choice*	2.5	[8]
CC($c\mathbb{R}$)		4.55	
CC(fin)		2.9	[10]
CC(n)		2.9	[288]
CC($\leq n$)		E 1.Sec. 3.1	[374]
CC($\mathbb{R}$)		2.9	[94]
CC($\mathbb{Z}$)		2.9 and Sec. 4.7	[119]
CC(2)		3.4	[80]
CH	*Continuum Hypothesis*	2.19	
CMC	*Axiom of Countable Multiple Choice*	2.10	[126]
CUT	*Countable Union Theorem*	3.2	[31]
CUT(fin)		3.2	[10]
CUT(n)		E 1.Sec.3.1	[374]

CUT($\leq n$)		E 1.Sec.3.1	
CUT($\mathbb{R}$)		3.2	[6]
CUT(2)		3.2	[80]
DC	*Principle of Dependent Choices*	2.11	[43]
DMC	*Principle of Dependent Multiple Choices*	E 2.Sec.2.2 and E 5.Sec.4.10	[106]
EAC	*Axiom of Even Choice*	E 1.Sec.2.1	
	Existence of a Hamel bases for $\mathbb{R}$	Section 5.1	[367]
	Existence of non–measurable sets	Section 5.1	[93]
	Existence of ugly functions	Section 5.1	[366]
Fin	*finite = D–finite*	2.13	[9]
Fin(lin)		2.13	[185]
Fin($\mathbb{R}$)		2.13	[13]
GCH	*Generalized Continuum Hypothesis*	2.19	
HBT	*Hahn–Banach Theorem*	5.25	[52]
	Hausdorff's Maximal Chain Condition	2.2	[1]
	Kurepa's Maximal Antichain Condition	2.4	[1]
KW	*Kinna–Wagner Selection Principle*	2.8	[15]
	Lebesgue–measure is σ*–additive*	E 3.Sec.5.1	[37]
	$m = 2m$ *for infinite cardinals*	E 7.Sec.4.1, E 3.Sec.4.2	[3]
	No amorphous sets exist	E 11.Sec.4.1	[64]
OAC	*Axiom of Odd Choice*	E 1.Sec.2.1	
OEP	*Order Extension Principle*	2.17	[49]
OP	*Ordering Principle*	2.17	[30]
$\omega -$ **CC**($\mathbb{R}$)		4.56	
PCC	*Axiom of Partial Countable Choice*	2.11	[8]
PCC(fin)		E 5.Sec.2.2	[10]
PCC($\mathbb{R}$)		E 5.Sec.2.2	[94]
PCC(2)		E 2.Sec.3.1	[373(2)]
PCMC		E 5.Sec.2.2	[126]
PIT	*Boolean Prime Ideal Theorem*	2.15	[14]
	$\mathbb{R}$ *is not a countable union of countable sets*	4.58	[38]
	$\mathbb{R} \cong \mathbb{C}$	5.1	[251]
	$\mathbb{R} \cong \mathbb{R} \oplus \mathbb{Q}$	5.10	[252]
	$\mathbb{R}$ *is sequential*	4.55	[74]
	Teichmüller–Tukey Lemma	2.2	[1]
UFT	*Ultrafilter Theorem*	2.15	[14]

Index

C

D

K

L

M

N

O

P

Z

Lecture Notes in Mathematics

For information about earlier volumes
please contact your bookseller or Springer
LNM Online archive: springerlink.com

Vol. 1674: G. Klaas, C. R. Leedham-Green, W. Plesken, Linear Pro-p-Groups of Finite Width (1997)
Vol. 1675: J. E. Yukich, Probability Theory of Classical Euclidean Optimization Problems (1998)
Vol. 1676: P. Cembranos, J. Mendoza, Banach Spaces of Vector-Valued Functions (1997)
Vol. 1677: N. Proskurin, Cubic Metaplectic Forms and Theta Functions (1998)
Vol. 1678: O. Krupková, The Geometry of Ordinary Variational Equations (1997)
Vol. 1679: K.-G. Grosse-Erdmann, The Blocking Technique. Weighted Mean Operators and Hardy's Inequality (1998)
Vol. 1680: K.-Z. Li, F. Oort, Moduli of Supersingular Abelian Varieties (1998)
Vol. 1681: G. J. Wirsching, The Dynamical System Generated by the 3n+1 Function (1998)
Vol. 1682: H.-D. Alber, Materials with Memory (1998)
Vol. 1683: A. Pomp, The Boundary-Domain Integral Method for Elliptic Systems (1998)
Vol. 1684: C. A. Berenstein, P. F. Ebenfelt, S. G. Gindikin, S. Helgason, A. E. Tumanov, Integral Geometry, Radon Transforms and Complex Analysis. Firenze, 1996. Editors: E. Casadio Tarabusi, M. A. Picardello, G. Zampieri (1998)
Vol. 1685: S. König, A. Zimmermann, Derived Equivalences for Group Rings (1998)
Vol. 1686: J. Azéma, M. Émery, M. Ledoux, M. Yor (Eds.), Séminaire de Probabilités XXXII (1998)
Vol. 1687: F. Bornemann, Homogenization in Time of Singularly Perturbed Mechanical Systems (1998)
Vol. 1688: S. Assing, W. Schmidt, Continuous Strong Markov Processes in Dimension One (1998)
Vol. 1689: W. Fulton, P. Pragacz, Schubert Varieties and Degeneracy Loci (1998)
Vol. 1690: M. T. Barlow, D. Nualart, Lectures on Probability Theory and Statistics. Editor: P. Bernard (1998)
Vol. 1691: R. Bezrukavnikov, M. Finkelberg, V. Schechtman, Factorizable Sheaves and Quantum Groups (1998)
Vol. 1692: T. M. W. Eyre, Quantum Stochastic Calculus and Representations of Lie Superalgebras (1998)
Vol. 1694: A. Braides, Approximation of Free-Discontinuity Problems (1998)
Vol. 1695: D. J. Hartfiel, Markov Set-Chains (1998)
Vol. 1696: E. Bouscaren (Ed.): Model Theory and Algebraic Geometry (1998)
Vol. 1697: B. Cockburn, C. Johnson, C.-W. Shu, E. Tadmor, Advanced Numerical Approximation of Nonlinear Hyperbolic Equations. Cetraro, Italy, 1997. Editor: A. Quarteroni (1998)
Vol. 1698: M. Bhattacharjee, D. Macpherson, R. G. Möller, P. Neumann, Notes on Infinite Permutation Groups (1998)
Vol. 1699: A. Inoue,Tomita-Takesaki Theory in Algebras of Unbounded Operators (1998)
Vol. 1700: W. A. Woyczyński, Burgers-KPZ Turbulence (1998)
Vol. 1701: Ti-Jun Xiao, J. Liang, The Cauchy Problem of Higher Order Abstract Differential Equations (1998)
Vol. 1702: J. Ma, J. Yong, Forward-Backward Stochastic Differential Equations and Their Applications (1999)
Vol. 1703: R. M. Dudley, R. Norvaiša, Differentiability of Six Operators on Nonsmooth Functions and p-Variation (1999)
Vol. 1704: H. Tamanoi, Elliptic Genera and Vertex Operator Super-Algebras (1999)
Vol. 1705: I. Nikolaev, E. Zhuzhoma, Flows in 2-dimensional Manifolds (1999)
Vol. 1706: S. Yu. Pilyugin, Shadowing in Dynamical Systems (1999)
Vol. 1707: R. Pytlak, Numerical Methods for Optimal Control Problems with State Constraints (1999)
Vol. 1708: K. Zuo, Representations of Fundamental Groups of Algebraic Varieties (1999)
Vol. 1709: J. Azéma, M. Émery, M. Ledoux, M. Yor (Eds.), Séminaire de Probabilités XXXIII (1999)
Vol. 1710: M. Koecher, The Minnesota Notes on Jordan Algebras and Their Applications (1999)
Vol. 1711: W. Ricker, Operator Algebras Generated by Commuting Projećtions: A Vector Measure Approach (1999)
Vol. 1712: N. Schwartz, J. J. Madden, Semi-algebraic Function Rings and Reflectors of Partially Ordered Rings (1999)
Vol. 1713: F. Bethuel, G. Huisken, S. Müller, K. Steffen, Calculus of Variations and Geometric Evolution Problems. Cetraro, 1996. Editors: S. Hildebrandt, M. Struwe (1999)
Vol. 1714: O. Diekmann, R. Durrett, K. P. Hadeler, P. K. Maini, H. L. Smith, Mathematics Inspired by Biology. Martina Franca, 1997. Editors: V. Capasso, O. Diekmann (1999)
Vol. 1715: N. V. Krylov, M. Röckner, J. Zabczyk, Stochastic PDE's and Kolmogorov Equations in Infinite Dimensions. Cetraro, 1998. Editor: G. Da Prato (1999)
Vol. 1716: J. Coates, R. Greenberg, K. A. Ribet, K. Rubin, Arithmetic Theory of Elliptic Curves. Cetraro, 1997. Editor: C. Viola (1999)
Vol. 1717: J. Bertoin, F. Martinelli, Y. Peres, Lectures on Probability Theory and Statistics. Saint-Flour, 1997. Editor: P. Bernard (1999)
Vol. 1718: A. Eberle, Uniqueness and Non-Uniqueness of Semigroups Generated by Singular Diffusion Operators (1999)
Vol. 1719: K. R. Meyer, Periodic Solutions of the N-Body Problem (1999)
Vol. 1720: D. Elworthy, Y. Le Jan, X-M. Li, On the Geometry of Diffusion Operators and Stochastic Flows (1999)
Vol. 1721: A. Iarrobino, V. Kanev, Power Sums, Gorenstein Algebras, and Determinantal Loci (1999)

Vol. 1722: R. McCutcheon, Elemental Methods in Ergodic Ramsey Theory (1999)
Vol. 1723: J. P. Croisille, C. Lebeau, Diffraction by an Immersed Elastic Wedge (1999)
Vol. 1724: V. N. Kolokoltsov, Semiclassical Analysis for Diffusions and Stochastic Processes (2000)
Vol. 1725: D. A. Wolf-Gladrow, Lattice-Gas Cellular Automata and Lattice Boltzmann Models (2000)
Vol. 1726: V. Marić, Regular Variation and Differential Equations (2000)
Vol. 1727: P. Kravanja M. Van Barel, Computing the Zeros of Analytic Functions (2000)
Vol. 1728: K. Gatermann Computer Algebra Methods for Equivariant Dynamical Systems (2000)
Vol. 1729: J. Azéma, M. Émery, M. Ledoux, M. Yor (Eds.) Séminaire de Probabilités XXXIV (2000)
Vol. 1730: S. Graf, H. Luschgy, Foundations of Quantization for Probability Distributions (2000)
Vol. 1731: T. Hsu, Quilts: Central Extensions, Braid Actions, and Finite Groups (2000)
Vol. 1732: K. Keller, Invariant Factors, Julia Equivalences and the (Abstract) Mandelbrot Set (2000)
Vol. 1733: K. Ritter, Average-Case Analysis of Numerical Problems (2000)
Vol. 1734: M. Espedal, A. Fasano, A. Mikelić, Filtration in Porous Media and Industrial Applications. Cetraro 1998. Editor: A. Fasano. 2000.
Vol. 1735: D. Yafaev, Scattering Theory: Some Old and New Problems (2000)
Vol. 1736: B. O. Turesson, Nonlinear Potential Theory and Weighted Sobolev Spaces (2000)
Vol. 1737: S. Wakabayashi, Classical Microlocal Analysis in the Space of Hyperfunctions (2000)
Vol. 1738: M. Émery, A. Nemirovski, D. Voiculescu, Lectures on Probability Theory and Statistics (2000)
Vol. 1739: R. Burkard, P. Deuflhard, A. Jameson, J.-L. Lions, G. Strang, Computational Mathematics Driven by Industrial Problems. Martina Franca, 1999. Editors: V. Capasso, H. Engl, J. Periaux (2000)
Vol. 1740: B. Kawohl, O. Pironneau, L. Tartar, J.-P. Zolesio, Optimal Shape Design. Tróia, Portugal 1999. Editors: A. Cellina, A. Ornelas (2000)
Vol. 1741: E. Lombardi, Oscillatory Integrals and Phenomena Beyond all Algebraic Orders (2000)
Vol. 1742: A. Unterberger, Quantization and Nonholomorphic Modular Forms (2000)
Vol. 1743: L. Habermann, Riemannian Metrics of Constant Mass and Moduli Spaces of Conformal Structures (2000)
Vol. 1744: M. Kunze, Non-Smooth Dynamical Systems (2000)
Vol. 1745: V. D. Milman, G. Schechtman (Eds.), Geometric Aspects of Functional Analysis. Israel Seminar 1999-2000 (2000)
Vol. 1746: A. Degtyarev, I. Itenberg, V. Kharlamov, Real Enriques Surfaces (2000)
Vol. 1747: L. W. Christensen, Gorenstein Dimensions (2000)
Vol. 1748: M. Ruzicka, Electrorheological Fluids: Modeling and Mathematical Theory (2001)
Vol. 1749: M. Fuchs, G. Seregin, Variational Methods for Problems from Plasticity Theory and for Generalized Newtonian Fluids (2001)
Vol. 1750: B. Conrad, Grothendieck Duality and Base Change (2001)
Vol. 1751: N. J. Cutland, Loeb Measures in Practice: Recent Advances (2001)
Vol. 1752: Y. V. Nesterenko, P. Philippon, Introduction to Algebraic Independence Theory (2001)
Vol. 1753: A. I. Bobenko, U. Eitner, Painlevé Equations in the Differential Geometry of Surfaces (2001)
Vol. 1754: W. Bertram, The Geometry of Jordan and Lie Structures (2001)
Vol. 1755: J. Azéma, M. Émery, M. Ledoux, M. Yor (Eds.), Séminaire de Probabilités XXXV (2001)
Vol. 1756: P. E. Zhidkov, Korteweg de Vries and Nonlinear Schrödinger Equations: Qualitative Theory (2001)
Vol. 1757: R. R. Phelps, Lectures on Choquet's Theorem (2001)
Vol. 1758: N. Monod, Continuous Bounded Cohomology of Locally Compact Groups (2001)
Vol. 1759: Y. Abe, K. Kopfermann, Toroidal Groups (2001)
Vol. 1760: D. Filipović, Consistency Problems for Heath-Jarrow-Morton Interest Rate Models (2001)
Vol. 1761: C. Adelmann, The Decomposition of Primes in Torsion Point Fields (2001)
Vol. 1762: S. Cerrai, Second Order PDE's in Finite and Infinite Dimension (2001)
Vol. 1763: J.-L. Loday, A. Frabetti, F. Chapoton, F. Goichot, Dialgebras and Related Operads (2001)
Vol. 1764: A. Cannas da Silva, Lectures on Symplectic Geometry (2001)
Vol. 1765: T. Kerler, V. V. Lyubashenko, Non-Semisimple Topological Quantum Field Theories for 3-Manifolds with Corners (2001)
Vol. 1766: H. Hennion, L. Hervé, Limit Theorems for Markov Chains and Stochastic Properties of Dynamical Systems by Quasi-Compactness (2001)
Vol. 1767: J. Xiao, Holomorphic Q Classes (2001)
Vol. 1768: M.J. Pflaum, Analytic and Geometric Study of Stratified Spaces (2001)
Vol. 1769: M. Alberich-Carramiñana, Geometry of the Plane Cremona Maps (2002)
Vol. 1770: H. Gluesing-Luerssen, Linear Delay-Differential Systems with Commensurate Delays: An Algebraic Approach (2002)
Vol. 1771: M. Émery, M. Yor (Eds.), Séminaire de Probabilités 1967-1980. A Selection in Martingale Theory (2002)
Vol. 1772: F. Burstall, D. Ferus, K. Leschke, F. Pedit, U. Pinkall, Conformal Geometry of Surfaces in S^4 (2002)
Vol. 1773: Z. Arad, M. Muzychuk, Standard Integral Table Algebras Generated by a Non-real Element of Small Degree (2002)
Vol. 1774: V. Runde, Lectures on Amenability (2002)
Vol. 1775: W. H. Meeks, A. Ros, H. Rosenberg, The Global Theory of Minimal Surfaces in Flat Spaces. Martina Franca 1999. Editor: G. P. Pirola (2002)
Vol. 1776: K. Behrend, C. Gomez, V. Tarasov, G. Tian, Quantum Comohology. Cetraro 1997. Editors: P. de Bartolomeis, B. Dubrovin, C. Reina (2002)
Vol. 1777: E. García-Río, D. N. Kupeli, R. Vázquez-Lorenzo, Osserman Manifolds in Semi-Riemannian Geometry (2002)
Vol. 1778: H. Kiechle, Theory of K-Loops (2002)
Vol. 1779: I. Chueshov, Monotone Random Systems (2002)
Vol. 1780: J. H. Bruinier, Borcherds Products on O(2,1) and Chern Classes of Heegner Divisors (2002)
Vol. 1781: E. Bolthausen, E. Perkins, A. van der Vaart, Lectures on Probability Theory and Statistics. Ecole d' Eté de Probabilités de Saint-Flour XXIX-1999. Editor: P. Bernard (2002)

Vol. 1782: C.-H. Chu, A. T.-M. Lau, Harmonic Functions on Groups and Fourier Algebras (2002)
Vol. 1783: L. Grüne, Asymptotic Behavior of Dynamical and Control Systems under Perturbation and Discretization (2002)
Vol. 1784: L.H. Eliasson, S. B. Kuksin, S. Marmi, J.-C. Yoccoz, Dynamical Systems and Small Divisors. Cetraro, Italy 1998. Editors: S. Marmi, J.-C. Yoccoz (2002)
Vol. 1785: J. Arias de Reyna, Pointwise Convergence of Fourier Series (2002)
Vol. 1786: S. D. Cutkosky, Monomialization of Morphisms from 3-Folds to Surfaces (2002)
Vol. 1787: S. Caenepeel, G. Militaru, S. Zhu, Frobenius and Separable Functors for Generalized Module Categories and Nonlinear Equations (2002)
Vol. 1788: A. Vasil'ev, Moduli of Families of Curves for Conformal and Quasiconformal Mappings (2002)
Vol. 1789: Y. Sommerhäuser, Yetter-Drinfel'd Hopf algebras over groups of prime order (2002)
Vol. 1790: X. Zhan, Matrix Inequalities (2002)
Vol. 1791: M. Knebusch, D. Zhang, Manis Valuations and Prüfer Extensions I: A new Chapter in Commutative Algebra (2002)
Vol. 1792: D. D. Ang, R. Gorenflo, V. K. Le, D. D. Trong, Moment Theory and Some Inverse Problems in Potential Theory and Heat Conduction (2002)
Vol. 1793: J. Cortés Monforte, Geometric, Control and Numerical Aspects of Nonholonomic Systems (2002)
Vol. 1794: N. Pytheas Fogg, Substitution in Dynamics, Arithmetics and Combinatorics. Editors: V. Berthé, S. Ferenczi, C. Mauduit, A. Siegel (2002)
Vol. 1795: H. Li, Filtered-Graded Transfer in Using Noncommutative Gröbner Bases (2002)
Vol. 1796: J.M. Melenk, hp-Finite Element Methods for Singular Perturbations (2002)
Vol. 1797: B. Schmidt, Characters and Cyclotomic Fields in Finite Geometry (2002)
Vol. 1798: W.M. Oliva, Geometric Mechanics (2002)
Vol. 1799: H. Pajot, Analytic Capacity, Rectifiability, Menger Curvature and the Cauchy Integral (2002)
Vol. 1800: O. Gabber, L. Ramero, Almost Ring Theory (2003)
Vol. 1801: J. Azéma, M. Émery, M. Ledoux, M. Yor (Eds.), Séminaire de Probabilités XXXVI (2003)
Vol. 1802: V. Capasso, E. Merzbach, B.G. Ivanoff, M. Dozzi, R. Dalang, T. Mountford, Topics in Spatial Stochastic Processes. Martina Franca, Italy 2001. Editor: E. Merzbach (2003)
Vol. 1803: G. Dolzmann, Variational Methods for Crystalline Microstructure – Analysis and Computation (2003)
Vol. 1804: I. Cherednik, Ya. Markov, R. Howe, G. Lusztig, Iwahori-Hecke Algebras and their Representation Theory. Martina Franca, Italy 1999. Editors: V. Baldoni, D. Barbasch (2003)
Vol. 1805: F. Cao, Geometric Curve Evolution and Image Processing (2003)
Vol. 1806: H. Broer, I. Hoveijn. G. Lunther, G. Vegter, Bifurcations in Hamiltonian Systems. Computing Singularities by Gröbner Bases (2003)
Vol. 1807: V. D. Milman, G. Schechtman (Eds.), Geometric Aspects of Functional Analysis. Israel Seminar 2000-2002 (2003)
Vol. 1808: W. Schindler, Measures with Symmetry Properties (2003)
Vol. 1809: O. Steinbach, Stability Estimates for Hybrid Coupled Domain Decomposition Methods (2003)
Vol. 1810: J. Wengenroth, Derived Functors in Functional Analysis (2003)
Vol. 1811: J. Stevens, Deformations of Singularities (2003)
Vol. 1812: L. Ambrosio, K. Deckelnick, G. Dziuk, M. Mimura, V. A. Solonnikov, H. M. Soner, Mathematical Aspects of Evolving Interfaces. Madeira, Funchal, Portugal 2000. Editors: P. Colli, J. F. Rodrigues (2003)
Vol. 1813: L. Ambrosio, L. A. Caffarelli, Y. Brenier, G. Buttazzo, C. Villani, Optimal Transportation and its Applications. Martina Franca, Italy 2001. Editors: L. A. Caffarelli, S. Salsa (2003)
Vol. 1814: P. Bank, F. Baudoin, H. Föllmer, L.C.G. Rogers, M. Soner, N. Touzi, Paris-Princeton Lectures on Mathematical Finance 2002 (2003)
Vol. 1815: A. M. Vershik (Ed.), Asymptotic Combinatorics with Applications to Mathematical Physics. St. Petersburg, Russia 2001 (2003)
Vol. 1816: S. Albeverio, W. Schachermayer, M. Talagrand, Lectures on Probability Theory and Statistics. Ecole d'Eté de Probabilités de Saint-Flour XXX-2000. Editor: P. Bernard (2003)
Vol. 1817: E. Koelink, W. Van Assche(Eds.), Orthogonal Polynomials and Special Functions. Leuven 2002 (2003)
Vol. 1818: M. Bildhauer, Convex Variational Problems with Linear, nearly Linear and/or Anisotropic Growth Conditions (2003)
Vol. 1819: D. Masser, Yu. V. Nesterenko, H. P. Schlickewei, W. M. Schmidt, M. Waldschmidt, Diophantine Approximation. Cetraro, Italy 2000. Editors: F. Amoroso, U. Zannier (2003)
Vol. 1820: F. Hiai, H. Kosaki, Means of Hilbert Space Operators (2003)
Vol. 1821: S. Teufel, Adiabatic Perturbation Theory in Quantum Dynamics (2003)
Vol. 1822: S.-N. Chow, R. Conti, R. Johnson, J. Mallet-Paret, R. Nussbaum, Dynamical Systems. Cetraro, Italy 2000. Editors: J. W. Macki, P. Zecca (2003)
Vol. 1823: A. M. Anile, W. Allegretto, C. Ringhofer, Mathematical Problems in Semiconductor Physics. Cetraro, Italy 1998. Editor: A. M. Anile (2003)
Vol. 1824: J. A. Navarro González, J. B. Sancho de Salas, $\mathscr{C}^\infty$ – Differentiable Spaces (2003)
Vol. 1825: J. H. Bramble, A. Cohen, W. Dahmen, Multiscale Problems and Methods in Numerical Simulations, Martina Franca, Italy 2001. Editor: C. Canuto (2003)
Vol. 1826: K. Dohmen, Improved Bonferroni Inequalities via Abstract Tubes. Inequalities and Identities of Inclusion-Exclusion Type. VIII, 113 p, 2003.
Vol. 1827: K. M. Pilgrim, Combinations of Complex Dynamical Systems. IX, 118 p, 2003.
Vol. 1828: D. J. Green, Gröbner Bases and the Computation of Group Cohomology. XII, 138 p, 2003.
Vol. 1829: E. Altman, B. Gaujal, A. Hordijk, Discrete-Event Control of Stochastic Networks: Multimodularity and Regularity. XIV, 313 p, 2003.
Vol. 1830: M. I. Gil', Operator Functions and Localization of Spectra. XIV, 256 p, 2003.
Vol. 1831: A. Connes, J. Cuntz, E. Guentner, N. Higson, J. E. Kaminker, Noncommutative Geometry, Martina Franca, Italy 2002. Editors: S. Doplicher, L. Longo (2004)
Vol. 1832: J. Azéma, M. Émery, M. Ledoux, M. Yor (Eds.), Séminaire de Probabilités XXXVII (2003)
Vol. 1833: D.-Q. Jiang, M. Qian, M.-P. Qian, Mathematical Theory of Nonequilibrium Steady States. On the Frontier of Probability and Dynamical Systems. IX, 280 p, 2004.

Vol. 1834: Yo. Yomdin, G. Comte, Tame Geometry with Application in Smooth Analysis. VIII, 186 p, 2004.
Vol. 1835: O.T. Izhboldin, B. Kahn, N.A. Karpenko, A. Vishik, Geometric Methods in the Algebraic Theory of Quadratic Forms. Summer School, Lens, 2000. Editor: J.-P. Tignol (2004)
Vol. 1836: C. Năstăsescu, F. Van Oystaeyen, Methods of Graded Rings. XIII, 304 p, 2004.
Vol. 1837: S. Tavaré, O. Zeitouni, Lectures on Probability Theory and Statistics. Ecole d'Eté de Probabilités de Saint-Flour XXXI-2001. Editor: J. Picard (2004)
Vol. 1838: A.J. Ganesh, N.W. O'Connell, D.J. Wischik, Big Queues. XII, 254 p, 2004.
Vol. 1839: R. Gohm, Noncommutative Stationary Processes. VIII, 170 p, 2004.
Vol. 1840: B. Tsirelson, W. Werner, Lectures on Probability Theory and Statistics. Ecole d'Eté de Probabilités de Saint-Flour XXXII-2002. Editor: J. Picard (2004)
Vol. 1841: W. Reichel, Uniqueness Theorems for Variational Problems by the Method of Transformation Groups (2004)
Vol. 1842: T. Johnsen, A.L. Knutsen, K3 Projective Models in Scrolls (2004)
Vol. 1843: B. Jefferies, Spectral Properties of Noncommuting Operators (2004)
Vol. 1844: K.F. Siburg, The Principle of Least Action in Geometry and Dynamics (2004)
Vol. 1845: Min Ho Lee, Mixed Automorphic Forms, Torus Bundles, and Jacobi Forms (2004)
Vol. 1846: H. Ammari, H. Kang, Reconstruction of Small Inhomogeneities from Boundary Measurements (2004)
Vol. 1847: T.R. Bielecki, T. Björk, M. Jeanblanc, M. Rutkowski, J.A. Scheinkman, W. Xiong, Paris-Princeton Lectures on Mathematical Finance 2003 (2004)
Vol. 1848: M. Abate, J. E. Fornaess, X. Huang, J. P. Rosay, A. Tumanov, Real Methods in Complex and CR Geometry, Martina Franca, Italy 2002. Editors: D. Zaitsev, G. Zampieri (2004)
Vol. 1849: Martin L. Brown, Heegner Modules and Elliptic Curves (2004)
Vol. 1850: V. D. Milman, G. Schechtman (Eds.), Geometric Aspects of Functional Analysis. Israel Seminar 2002-2003 (2004)
Vol. 1851: O. Catoni, Statistical Learning Theory and Stochastic Optimization (2004)
Vol. 1852: A.S. Kechris, B.D. Miller, Topics in Orbit Equivalence (2004)
Vol. 1853: Ch. Favre, M. Jonsson, The Valuative Tree (2004)
Vol. 1854: O. Saeki, Topology of Singular Fibers of Differential Maps (2004)
Vol. 1855: G. Da Prato, P.C. Kunstmann, I. Lasiecka, A. Lunardi, R. Schnaubelt, L. Weis, Functional Analytic Methods for Evolution Equations. Editors: M. Iannelli, R. Nagel, S. Piazzera (2004)
Vol. 1856: K. Back, T.R. Bielecki, C. Hipp, S. Peng, W. Schachermayer, Stochastic Methods in Finance, Bressanone/Brixen, Italy, 2003. Editors: M. Fritelli, W. Runggaldier (2004)
Vol. 1857: M. Émery, M. Ledoux, M. Yor (Eds.), Séminaire de Probabilités XXXVIII (2005)
Vol. 1858: A.S. Cherny, H.-J. Engelbert, Singular Stochastic Differential Equations (2005)
Vol. 1859: E. Letellier, Fourier Transforms of Invariant Functions on Finite Reductive Lie Algebras (2005)
Vol. 1860: A. Borisyuk, G.B. Ermentrout, A. Friedman, D. Terman, Tutorials in Mathematical Biosciences I. Mathematical Neurosciences (2005)
Vol. 1861: G. Benettin, J. Henrard, S. Kuksin, Hamiltonian Dynamics – Theory and Applications, Cetraro, Italy, 1999. Editor: A. Giorgilli (2005)
Vol. 1862: B. Helffer, F. Nier, Hypoelliptic Estimates and Spectral Theory for Fokker-Planck Operators and Witten Laplacians (2005)
Vol. 1863: H. Fürh, Abstract Harmonic Analysis of Continuous Wavelet Transforms (2005)
Vol. 1864: K. Efstathiou, Metamorphoses of Hamiltonian Systems with Symmetries (2005)
Vol. 1865: D. Applebaum, B.V. R. Bhat, J. Kustermans, J. M. Lindsay, Quantum Independent Increment Processes I. From Classical Probability to Quantum Stochastic Calculus. Editors: M. Schürmann, U. Franz (2005)
Vol. 1866: O.E. Barndorff-Nielsen, U. Franz, R. Gohm, B. Kümmerer, S. Thorbjønsen, Quantum Independent Increment Processes II. Structure of Quantum Lévy Processes, Classical Probability, and Physics. Editors: M. Schürmann, U. Franz, (2005)
Vol. 1867: J. Sneyd (Ed.), Tutorials in Mathematical Biosciences II. Mathematical Modeling of Calcium Dynamics and Signal Transduction. (2005)
Vol. 1868: J. Jorgenson, S. Lang, $Pos_n(R)$ and Eisenstein Sereies. (2005)
Vol. 1869: A. Dembo, T. Funaki, Lectures on Probability Theory and Statistics. Ecole d'Eté de Probabilités de Saint-Flour XXXIII-2003. Editor: J. Picard (2005)
Vol. 1870: V.I. Gurariy, W. Lusky, Geometry of Müntz Spaces and Related Questions. (2005)
Vol. 1871: P. Constantin, G. Gallavotti, A.V. Kazhikhov, Y. Meyer, S. Ukai, Mathematical Foundation of Turbulent Viscous Flows, Martina Franca, Italy, 2003. Editors: M. Cannone, T. Miyakawa (2006)
Vol. 1872: A. Friedman (Ed.), Tutorials in Mathematical Biosciences III. Cell Cycle, Proliferation, and Cancer (2006)
Vol. 1873: R. Mansuy, M. Yor, Random Times and Enlargements of Filtrations in a Brownian Setting (2006)
Vol. 1875: J. Pitman, J. Picard, Combinatorial Stochastic Processes. Ecole d'Eté de Probabilités de Saint-Flour XXXII-2002. Editor: J. Picard (2006)
Vol. 1876: H. Herrlich, Axiom of Choice. (2006)

Recent Reprints and New Editions

Vol. 1200: V. D. Milman, G. Schechtman (Eds.), Asymptotic Theory of Finite Dimensional Normed Spaces. 1986. – Corrected Second Printing (2001)
Vol. 1471: M. Courtieu, A.A. Panchishkin, Non-Archimedean L-Functions and Arithmetical Siegel Modular Forms. – Second Edition (2003)
Vol. 1618: G. Pisier, Similarity Problems and Completely Bounded Maps. 1995 – Second, Expanded Edition (2001)
Vol. 1629: J.D. Moore, Lectures on Seiberg-Witten Invariants. 1997 – Second Edition (2001)
Vol. 1638: P. Vanhaecke, Integrable Systems in the realm of Algebraic Geometry. 1996 – Second Edition (2001)
Vol. 1702: J. Ma, J. Yong, Forward-Backward Stochastic Differential Equations and their Applications. 1999. – Corrected 3rd printing (2005)